Digital Electronics - Logic Gates

FIRST EDITION

BY PRASUN BARUA

ABOUT

Welcome to Digital Electronics - Logic Gates! This is a nonfiction science book which contains various topics on logic gates of digital electronics. A logic gate is a device that serves as a foundation for digital circuits. They carry out fundamental logical functions in digital circuits. Most electronic devices we use today contain logic gates of Some kind. Logic gates, for example, can be found in technology such as smartphones, tablets, and memory devices. Logic gates in a circuit make judgments depending on a mix of digital signals from their inputs. The majority of logic gates have two inputs and one output. Boolean algebra is the foundation of logic gates. Every terminal is in one of two binary states at any one time: false or true. True represents 1 and false represents 0. The binary output will vary depending on the type of logic gate being utilized and the mix of inputs. A logic gate can be compared to a light switch, with the output being 0 in one position and 1 in the other. Integrated circuits typically employ logic gates (IC). This book contains various topics such as Digital Logic Gates, Logic AND Gates, Logic OR Gate, Logic NOT Gate, Logic NAND Gate, Logic NOR Gate, Exclusive-OR Gate, Exclusive-NOR Gate, Exclusive-OR Gate, Digital Logic Gates in A Brief, Pull-Up Resistors and Universal Logic Gates. This is the first edition of the book. Thanks for reading the book.

TABLE OF CONTENTS

A digital logic gate can have more than one input, for example, inputs A, B, C, D etc., but generally only have one digital output, (Q). Individual logic gates can be connected or cascaded together to form a logic gate function with any desired number of inputs, or to form combinational and sequential type circuits, or to produce different logic gate functions from standard gates.

Standard commercially available digital logic gates are available in two basic families or forms, **TTL** which stands for *Transistor-Transistor Logic* such as the 7400 series, and **CMOS** which stands for *Complementary Metal-Oxide-Silicon* which is the 4000 series of chips. This notation of TTL or CMOS refers to the logic technology used to manufacture the integrated circuit, (IC) or a "chip" as it is more Normally called.

Digital Logic Gate

Basically, **TTL** logic ICs use NPN and PNP type Bipolar Junction Transistors while **CMOS** logic ICs use complementary MOSFET or JFET type Field Effect Transistors for both their input and output circuitry.

Same like TTL and CMOS technology, simple digital logic gates can Also be made by connecting together diodes, transistors and resistors to produce RTL, Resistor-

Transistor logic gates, DTL, Diode-Transistor logic gates or ECL, Emitter-Coupled logic gates but these are less common now compared to the popular CMOS family.

Integrated Circuits or ICs as they are more Normally called, can be grouped together into families according to the number of transistors or "gates" that they contain. For example, a simple AND gate my contain only a few individual transistors, were as a more complex microprocessor may contain many thousands of individual transistor gates. Integrated circuits are categorized according to the number of logic gates or the complexity of the circuits within a single chip with the general classification for the number of individual gates given as:

Classification of Integrated Circuits

- Small Scale Integration or (SSI) – Contain up to 10 transistors or a few gates within a single package such as AND, OR, NOT gates.
- Medium Scale Integration or (MSI) – between 10 and 100 transistors or tens of gates within a single package and perform digital operations such as adders, decoders, counters, flip-flops and multiplexers.
- Large Scale Integration or (LSI) – between 100 and 1,000 transistors or hundreds of gates and perform specific digital operations such as I/O chips, memory, arithmetic and logic units.
- Very-Large Scale Integration or (VLSI) – between 1,000 and 10,000 transistors or thousands of gates and perform computational operations such as processors, large memory arrays and programmable logic devices.
- Super-Large-Scale Integration or (SLSI) – between 10,000 and 100,000 transistors within a single package and perform computational operations such as microprocessor chips, micro-controllers, basic PICs and calculators.
- Ultra-Large-Scale Integration or (ULSI) – more than 1 million transistors – the big boys that are used in computers CPUs, GPUs, video processors, micro-controllers, FPGAs and complex PICs.

While the "ultra large scale" ULSI classification is less well used, another level of integration which represents the complexity of the Integrated Circuit is known as the System-on-Chip or (**THEREFOREC**) for short. Here the individual components such as the microprocessor, memory, peripherals, I/O logic etc., are all produced on a single piece of silicon and which represents a whole electronic system within one single chip, literally putting the word "integrated" into integrated circuit.

These complete integrated chips which can contain up to 100 million individual silicon-CMOS transistor gates within one single package are generally used in mobile phones, digital cameras, micro-controllers, PICs and robotic type applications.

Moore's Law

Gordon Moore co-founder of the Intel corporation predicted that "The number of transistors and resistors on a single chip will double every 18 months" regarding the development of semiconductor gate technology in 1965. When Gordon Moore made his famous comment way back in 1965 there were approximately only 60 individual transistor gates on a single silicon chip or die.

First microprocessor of the world in 1971 was the Intel 4004 that had a 4-bit data bus and contained about 2,300 transistors on a single chip, operating at about 600kHz. Today, the Intel Corporation have placed a staggering 1.2 billion individual transistor gates onto its new Quad-core i7-2700K Sandy Bridge 64-bit microprocessor chip operating at nearly 4GHz, and the on-chip transistor count is still rising, as newer faster microprocessors and micro-controllers are developed.

Digital Logic States

The **Digital Logic Gate** is the basic building block from which all digital electronic circuits and microprocessor-based systems are constructed from. Basic digital logic gates perform logical operations of AND, OR and NOT on binary numbers.

In digital logic design only two voltage levels or states are allowed and these states are generally referred to as Logic "1" and Logic "0", or HIGH and LOW, or TRUE and FALSE. These two states are represented in Boolean Algebra and standard truth tables by the binary digits of "**1**" and "**0**" respectively.

A good example of a digital state is a simple light switch. The switch can be either "ON" or "OFF", one state or the other, but not both at the same time. Then we can summarize the relationship between these various digital states as being:

Boolean Algebra Boolean Logic Voltage State
Logic "1"   TRUE (T)   HIGH (H)
Logic "0"   FALSE (F)   LOW (L)

Most *digital logic gates* and digital logic systems use "Positive logic", in which a logic level "0" or "LOW" is represented by a zero voltage, 0v or ground and a logic level "1" or "HIGH" is represented by a higher voltage such as +5 volts, with the switching from one voltage level to the other, from either a logic level "0" to a "1" or a "1" to a "0" being made as quickly as possible to prevent any faulty operation of the logic circuit.

There Also exists a complementary "Negative Logic" system in which the values and the rules of a logic "0" and a logic "1" are reversed but in this chapter section about digital logic gates we shall only refer to the positive logic convention as it is the most Normally used.

In standard TTL (transistor-transistor logic) ICs there is a pre-defined voltage range for the input and output voltage levels which define exactly what is a logic "1" level and what is a logic "0" level and these are shown below.

TTL Input & Output Voltage Levels

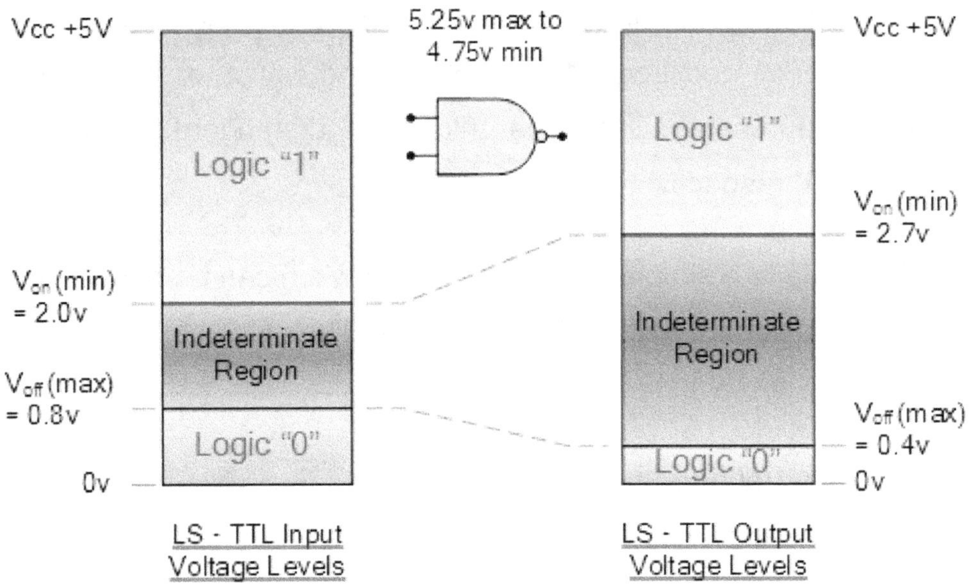

There are a large variety of logic gate types in both the bipolar 7400 and the CMOS 4000 families of digital logic gates such as 74Lxx, 74LSxx, 74ALSxx, 74HCxx, 74HCTxx, 74ACTxx etc., with each one having its own distinct advantages and disadvantages compared to the other. The exact switching voltage required to produce either a logic "0" or a logic "1" depends upon the specific logic group or family.

However, when using a standard +5 volt supply any TTL voltage input between 2.0v and 5v is considered to be a logic "1" or "HIGH" while any voltage input below 0.8v is recognized as a logic "0" or "LOW". The voltage region in between these two voltage levels either as an input or as an output is called the *Indeterminate Region* and operating within this region may cause the logic gate to produce a false output.

The CMOS 4000 logic family uses different levels of voltages compared to the TTL types as they are designed using field effect transistors, or FET's. In CMOS technology a logic "1" level operates between 3.0 and 18 volts and a logic "0" level is below 1.5 volts. Then the following table shows the difference between the logic levels of traditional TTL and CMOS logic gates.

TTL and CMOS Logic Levels

Device Type	Logic 0	Logic 1
TTL	0 to 0.8v	2.0 to 5v (V_{CC})
CMOS	0 to 1.5v	3.0 to 18v (V_{DD})

Then from the above observations, we can define the ideal TTL digital logic gate as one that has a "LOW" level logic "0" of 0 volts (ground) and a "HIGH" level logic "1" of +5 volts and this can be demonstrated as:

Ideal TTL Digital Logic Gate Voltage Levels

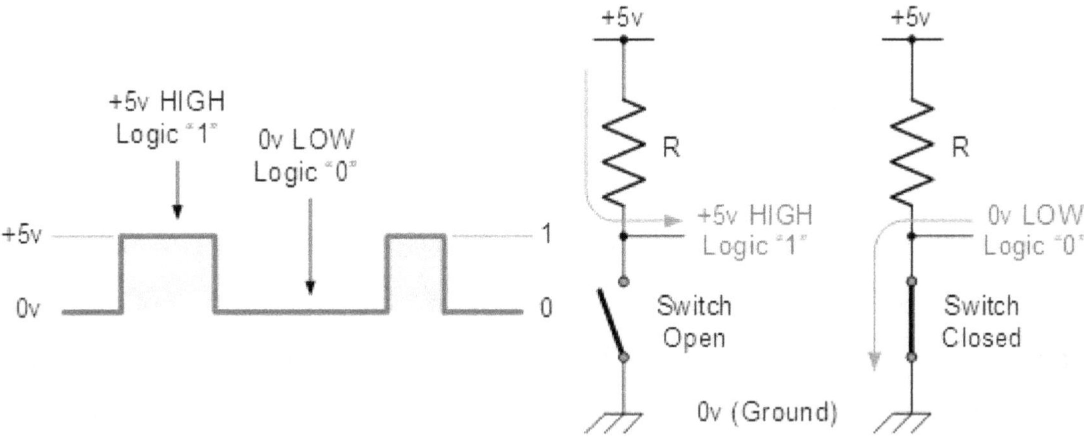

Where the opening or closing of the switch produces either a logic level "1" or a logic level "0" with the resistor R being known as a "pull-up" resistor.

Digital Logic Noise

However, between these defined HIGH and LOW values lies what is generally called a "no-man's land" (the blue area's above) and if we apply a signal voltage of a value within this no-man's land area we do not know whether the logic gate will respond to it as a level "0" or as a level "1", and the output will become unpredictable.

Noise is the name given to a random and unwanted voltage that is induced into electronic circuits by external interference, such as from nearby switches, power supply fluctuations or from wires and other conductors that pick-up stray electromagnetic radiation. Then in order for a logic gate not to be influence by noise in must have a certain amount of noise margin or noise immunity.

Digital Logic Gate Noise Immunity

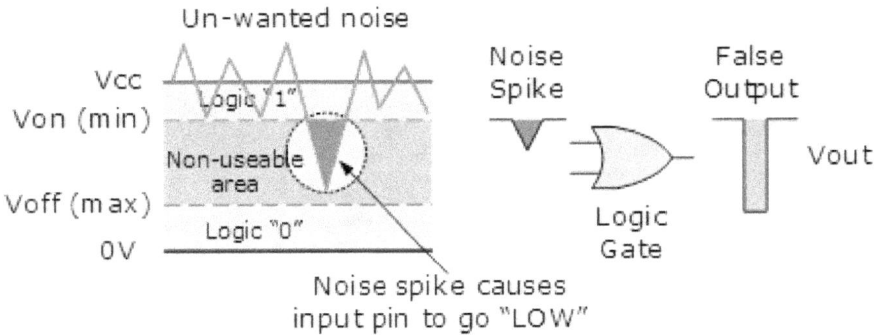

In the example above, the noise signal is superimposed onto the Vcc supply voltage and as long as it stays above the minimum level ($V_{ON(min)}$) the input an corresponding output of the logic gate are unaffected. But when the noise level becomes large enough and a noise spike causes the HIGH voltage level to drop below this minimum level, the logic gate may interpret this spike as a LOW-level input and switch the output accordingly producing a false output switching. Then in order for the logic gate not to be affected by noise it must be able to tolerate a certain amount of unwanted noise on its input without changing the state of its output.

Simple Basic Digital Logic Gates

Simple digital logic gates can be made by combining transistors, diodes and resistors with a simple example of a Diode-Resistor Logic (DRL) AND gate and a Diode-Transistor Logic (DTL) NAND gate given below.

Diode-Resistor Circuit

Diode-Transistor circuit

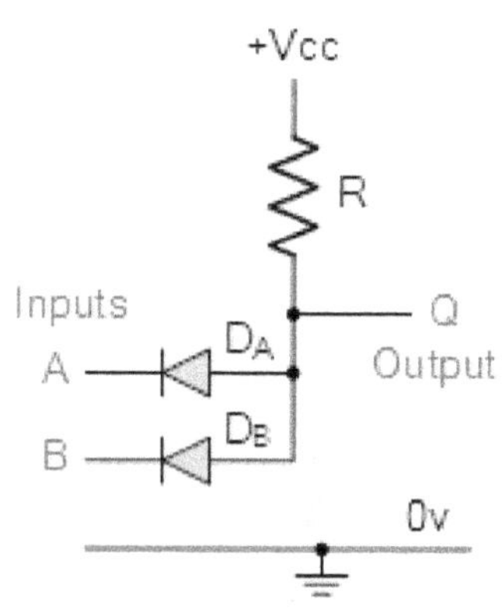

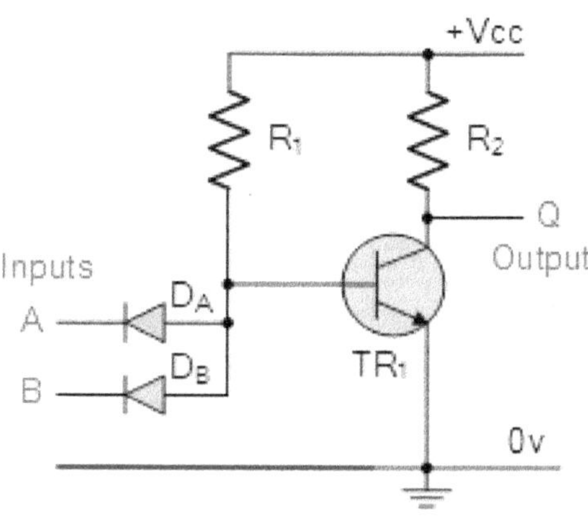

2-input AND Gate

2-input NAND Gate

The simple 2-input Diode-Resistor AND gate can be converted into a NAND gate by the addition of a single transistor inverting (NOT) stage. Using discrete components such as diodes, resistors and transistors to make digital logic gate circuits are not used in practical commercially available logic ICs as these circuits suffer from propagation delay or gate delay and Also power loss due to the pull-up resistors.

Another disadvantage of diode-resistor logic is that there is no "Fan-out" facility which is the ability of a single output to drive many inputs of the next stages. Also, this type of design does not turn fully "OFF" as a Logic "0" produces an output voltage of 0.6v (diode voltage drop), Therefore the following TTL and CMOS circuit designs are used instead.

Basic TTL Logic Gates

The simple Diode-Resistor AND gate above uses separate diodes for its inputs, one for each input. As a bipolar transistor is effectively two diode junctions connected together, representing either an NPN (Negative-Positive-Negative) device or a PNP (Positive-Negative-Positive) device, the input diodes of the diode-transistor logic (DTL) circuit can be replaced by one single NPN transistor with multiple emitter inputs to form another type of logic circuit called *transistor-transistor logic* or TTL as shown.

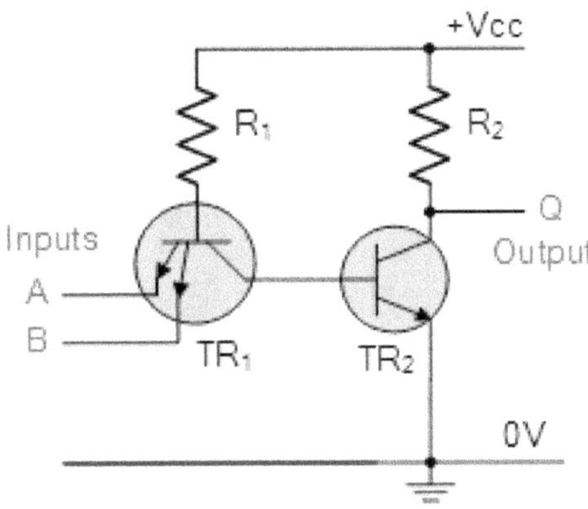

2-input NAND Gate

This simplified NAND gate circuit consists of an input transistor, TR_1 which has two (or more) emitter terminals and a single stage inverting NPN switching transistor circuit of TR_2.

When either or both of the emitters of TR_1 representing inputs "A" and "B" are connected to logic level "0" (LOW), the base current of TR_1 passes through its base/emitter junction to ground (0V), TR_1 saturates and its collector terminal follows. This action results in the base of TR_2 becoming connected to ground (0V), hence TR_2 is "OFF" and the output at Q is HIGH.

With both inputs "A" and "B" HIGH at logic level "1", input transistor TR_1 turns "OFF", the base of switching transistor TR_2 becomes HIGH and turns it "ON" Therefore the output at Q is LOW due to the switching action of the transistor. The multiple emitters of TR_1 are connected as inputs thus producing a NAND gate function.

Emitter-Coupled Digital Logic Gate

Emitter Coupled Logic or simply ECL, is another type of digital logic gate that uses bipolar transistor logic where the transistors are not operated in the saturation region, as they are with the standard TTL digital logic gate. Instead, the input and output circuits are push-pull connected transistors with the supply voltage negative with respect to ground.

This has the effect of increasing the speed of operation of the emitter coupled logic gates up to the Gigahertz range compared with the standard TTL types, but noise has a greater effect in ECL logic, as the unsaturated transistors operate within their active region and amplify as well as switch signals.

The "74" Sub-families of Integrated Circuits

With development in the circuit design to take account of propagation delays, current consumption, fan-in and fan-out requirements etc., this type of TTL bipolar transistor technology forms the basis of the prefixed "74" family of digital logic ICs, such as the "7400" Quad 2-input NAND gate, or the "7402" Quad 2-input NOR gate, etc.

Sub-families of the 74xxx series ICs are available relating to the different technologies used to fabricate the gates and they are denoted by the letters in between the 74 designation and the device number. There are a number of TTL sub-families available that provide a wide range of switching speeds and power consumption such as the 74L00 or 74**ALS**00 NAND gate, were the "L" stands for "Low-power TTL" and the "ALS" stands for "Advanced Low-power Schottky TTL" and these are listed below.

- 74xx or 74Nxx: Standard TTL – These devices are the original TTL family of logic gates introduced in the early 70's. They have a propagation delay of about 10ns and a power consumption of about 10mW. Supply voltage range: 4.75 to 5.25 volts
- 74Lxx: Low Power TTL – Power consumption was improved over standard types by increasing the number of internal resistances but at the cost of a reduction in switching speed. Supply voltage range: 4.75 to 5.25 volts
- 74Hxx: High Speed TTL – Switching speed was improved by reducing the number of internal resistances. This Also increased the power consumption. Supply voltage range: 4.75 to 5.25 volts

- 74Sxx: Schottky TTL – Schottky technology is used to improve input impedance, switching speed and power consumption (2mW) compared to the 74Lxx and 74Hxx types. Supply voltage range: 4.75 to 5.25 volts
- 74LSxx: Low Power Schottky TTL – Same as 74Sxx types but with increased internal resistances to improve power consumption. Supply voltage range: 4.75 to 5.25 volts
- 74ASxx: Advanced Schottky TTL – Improved design over 74Sxx Schottky types optimized to increase switching speed at the expense of power consumption of about 22mW. Supply voltage range: 4.5 to 5.5 volts
- 74ALSxx: Advanced Low Power Schottky TTL – Lower power consumption of about 1mW and higher switching speed of 4nS compared to 74LSxx types. Supply voltage range: 4.5 to 5.5 volts
- 74HCxx: High Speed CMOS – CMOS technology and transistors to reduce power consumption of less than 1uA with CMOS compatible inputs. Supply voltage range: 4.5 to 5.5 volts
- 74HCTxx: High Speed CMOS – CMOS technology and transistors to reduce power consumption of less than 1uA but has increased propagation delay of about 16nS due to the TTL compatible inputs. Supply voltage range: 4.5 to 5.5 volts

Basic CMOS Digital Logic Gate

One of the main disadvantages with the TTL digital logic gate series is that the logic gates are based on bipolar transistor logic technology and as transistors are current operated devices, they consume large amounts of power from a fixed +5 volt power supply.

Also, TTL bipolar transistor gates have a limited operating speed when switching from an "OFF" state to an "ON" state and vice-versa called the "gate" or "propagation delay". To overcome these limitations complementary MOS called "CMOS" (**C**omplementary **M**etal **O**xide **S**emiconductor) logic gates which use "Field Effect Transistors" or FET's were developed.

As these gates use both P-channel and N-channel MOSFET's as their input device, at quiescent conditions with no switching, the power consumption of CMOS gates is almost zero, (1 to 2µA) making them ideal for use in low-power battery circuits and with switching speeds upwards of 100MHz for use in high frequency timing and computer circuits.

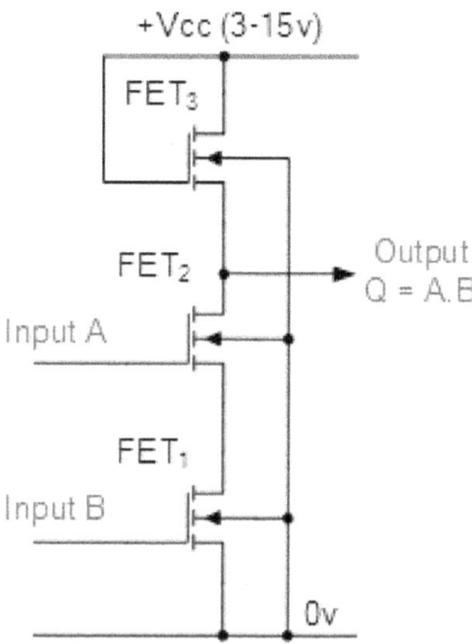

2-input NAND Gate

This basic CMOS gate example contains three N-channel normally-off enhancement MOSFET's, one for each input consisting of FET_1 and FET_2, and an additional switching MOSFET, FET_3 which is biased permanently "ON" through its gate.

When one or both inputs "A" and "B" are grounded to logic level "0", the corresponding input MOSFET, FET_1 or FET_2 are switched "OFF" producing a logic "1" (HIGH) output condition from the Source terminal of FET_3.

Only when both inputs "A" and "B" are held HIGH at logic level "1", does current flow through the corresponding MOSFET switching it "ON" producing an output state at Q equivalent to a logic level "0" as both MOSFETS, FET_1 and FET_2 are conducting. Therefore, producing the switching action representative of a NAND gate function.

Improvements in the circuit design with regards to switching speed, low power consumption and improved propagation delays has resulted in the standard CMOS 4000 "CD" family of logic ICs being developed that complement the TTL range.

As with the standard TTL digital logic gates, all the major digital logic gates and devices are available in the CMOS package such as the CD4011, a Quad 2-input NAND gate, or the CD4001, a Quad 2-input NOR gate along with all their sub-families.

Like TTL logic, complementary MOS (CMOS) circuits take advantage of the fact that both N-channel and P-channel devices can be fabricated together on the same substrate material to form various logic functions.

One of the main disadvantages with the CMOS range of ICs compared to their equivalent TTL types is that they are easily damaged by static electricity. Also, unlike TTL logic gates that operate on single +5V voltages for both their input and output levels, CMOS digital logic gates operate on a single supply voltage of between +3 and +18 volts.

Common CMOS Sub-families include:

- 4000B Series: Standard CMOS – These devices are the original Buffered CMOS family of logic gates introduced in the early 70's and operate from a supply voltage of 3.0 to 18v DC.
- 74C Series: 5v CMOS – These devices are pin-compatible with standard 5v TTL devices as their logic switching is implemented in CMOS but with TTL-compatible inputs. They operate from a supply voltage of 3.0 to 18v DC.

Note that CMOS logic gates and devices are static sensitive, therefore always take the appropriate precautions of working on antistatic mats or grounded workbenches, wearing an antistatic wristband and not removing a part from its antistatic packaging until required.

The Logic AND Gate is a type of digital logic circuit whose output goes HIGH to a logic level 1 only when all of its inputs are HIGH

The output state of a digital logic AND gate only returns "LOW" again when **ANY** of its inputs are at a logic level "0". In other words, for a logic AND gate, any LOW input will give a LOW output.

The logic or Boolean expression given for a digital AND gate is that for *Logical Multiplication* which is denoted by a single dot or full stop symbol, (.) giving us the Boolean expression of: A.B = Q.

Then we can define the operation of a digital 2-input AND gate as being:

"If both A and B are true, then Q is true"

2-input Transistor AND Gate

A simple 2-input AND gate can be constructed using RTL Resistor-transistor switches connected together as shown below with the inputs connected directly to the transistor bases. Both transistors must be saturated "ON" for an output at Q.

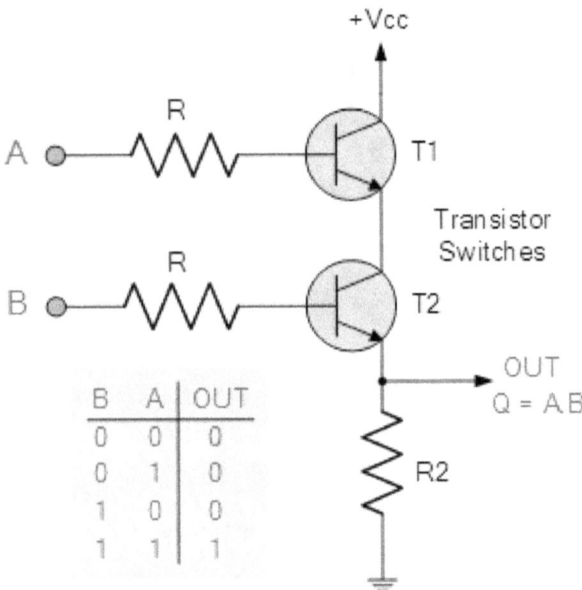

Logic AND Gates are available using digital circuits to produce the desired logical function and is given a symbol whose shape represents the logical operation of the AND gate.

Digital AND Gate Types

The 2-input Logic AND Gate

Symbol

2-input AND Gate

Truth Table

B	A	Q
0	0	0
0	1	0
1	0	0
1	1	1

Boolean Expression $Q = A.B$ Read as A **AND** B gives Q

The 3-input Logic AND Gate

Symbol   Truth Table

C	B	A	Q
0	0	0	0
0	0	1	0
0	1	0	0
0	1	1	0
1	0	0	0
1	0	1	0
1	1	0	0
1	1	1	1

3-input AND Gate

Boolean Expression $Q = A.B.C$ Read as A **AND** B **AND** C gives Q

Since the Boolean expression for the AND function is defined as (.), which is a binary operation, AND gates can be cascaded together to form any number of individual inputs. However, commercially available AND gate ICs are only available in standard 2, 3, or 4-input packages. If additional inputs are required, then standard AND gates will need to be cascaded together to obtain the required input value, for example.

Multi-input AND Gate

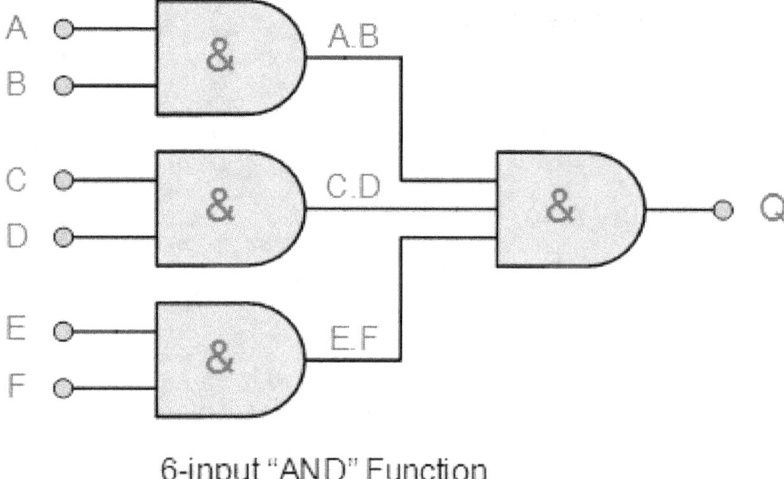

6-input "AND" Function

The Boolean Expression for this 6-input AND gate will therefore be:

Q = (A.B).(C.D).(E.F)

In other words:

A AND B AND C AND D AND E AND F gives Q

If the number of inputs required is an odd number of inputs any "unused" inputs can be held HIGH by connecting them directly to the power supply using suitable "Pull-up" resistors.

Normally available digital logic AND gate ICs include:

TTL Logic AND Gate

- 74LS08 Quad 2-input
- 74LS11 Triple 3-input
- 74LS21 Dual 4-input

CMOS Logic AND Gate

- CD4081 Quad 2-input
- CD4073 Triple 3-input
- CD4082 Dual 4-input

7408 Quad 2-input AND Gate

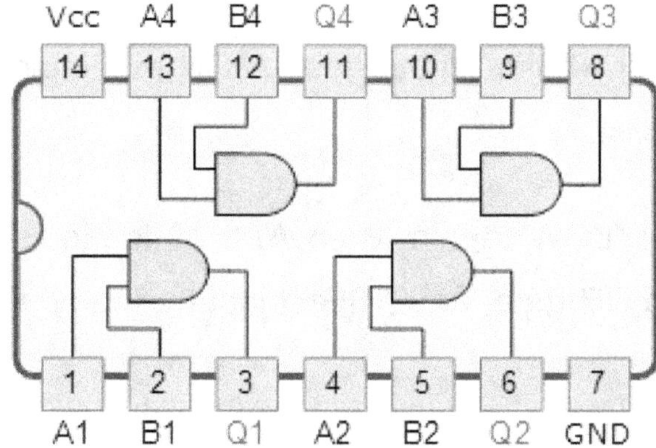

The Logic OR Gate is a type of digital logic circuit whose output goes HIGH to a logic level 1 only when one or more of its inputs are HIGH

The output, Q of a "Logic OR Gate" only returns "LOW" again when **ALL** of its inputs are at a logic level "0". In other words, for a logic OR gate, any "HIGH" input will give a "HIGH", logic level "1" output.

The logic or Boolean expression given for a digital logic OR gate is that for *Logical Addition* which is denoted by a plus sign, (+) giving us the Boolean expression of: A+B = Q.

Thus, the OR gate can be correctly described as an "Inclusive OR gate" As the output is true when both of its inputs are true (HIGH). Then we can define the operation of a 2-input logic OR gate as being:

"If either A or B is true, then Q is true"

2-input Transistor OR Gate

A simple 2-input inclusive OR gate can be constructed using RTL Resistor-transistor switches connected together as shown below with the inputs connected directly to the transistor bases. Either transistor must be saturated "ON" for an output at Q.

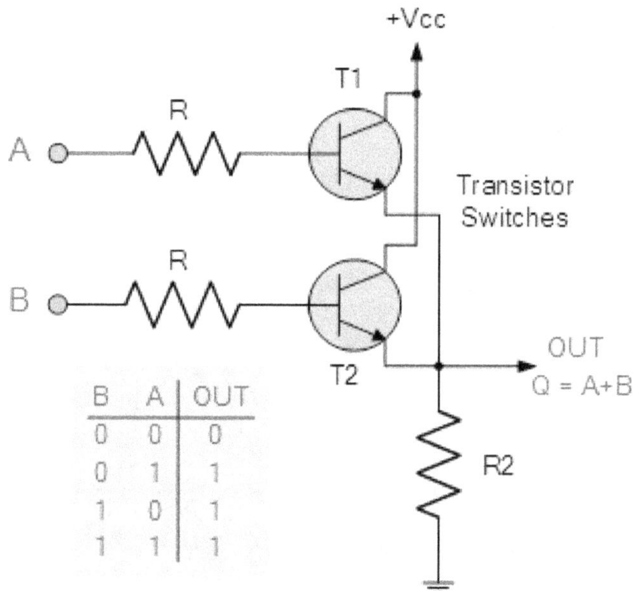

Logic OR Gates are available using digital circuits to produce the desired logical function and is given a symbol whose shape represents the logical operation of OR.

Digital Logic "OR" Gate Types

The 2-input Logic OR Gate

Symbol

2-input OR Gate

Truth Table

B	A	Q
0	0	0
0	1	1
1	0	1
1	1	1

Boolean Expression **Q = A+B**   Read as A **OR** B gives Q

The 3-input Logic OR Gate

Symbol   Truth Table

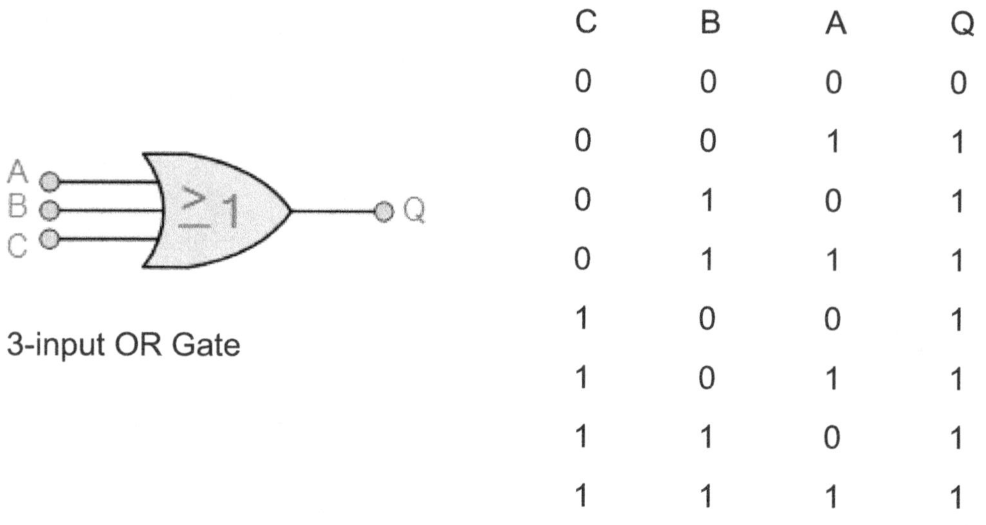

3-input OR Gate

C	B	A	Q
0	0	0	0
0	0	1	1
0	1	0	1
0	1	1	1
1	0	0	1
1	0	1	1
1	1	0	1
1	1	1	1

Boolean Expression **Q = A+B+C**   Read as A **OR** B **OR** C gives Q

Like the AND gate, the OR function can have any number of individual inputs. However, commercially available OR gates are available in 2, 3, or 4 inputs types. Additional inputs will require gates to be cascaded together for example.

Multi-input OR Gate

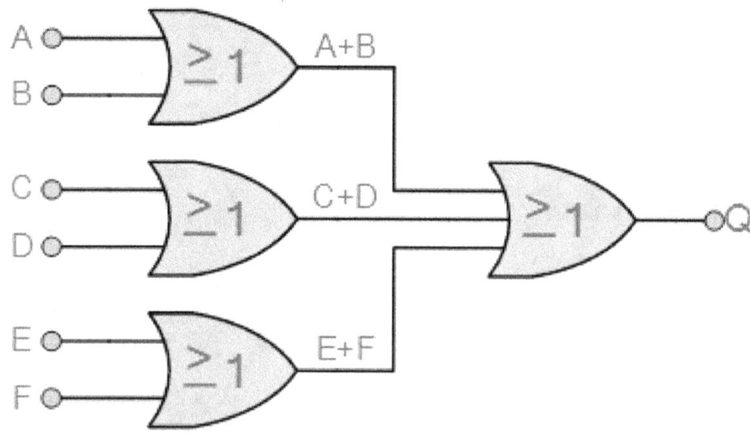

6-input "OR" Function

The Boolean Expression for this 6-input OR gate will therefore be:

Q = (A+B)+(C+D)+(E+F)

In other words:

A OR B OR C OR D OR E OR F gives Q

If the number of inputs required is an odd number of inputs any "unused" inputs can be held LOW by connecting them directly to ground using suitable "Pull-down" resistors.

Normally available digital logic OR gate ICs include:

<u>TTL Logic OR Gates</u>

- 74LS32 Quad 2-input

<u>CMOS Logic OR Gates</u>

- CD4071 Quad 2-input
- CD4075 Triple 3-input
- CD4072 Dual 4-input

7432 Quad 2-input Logic OR Gate

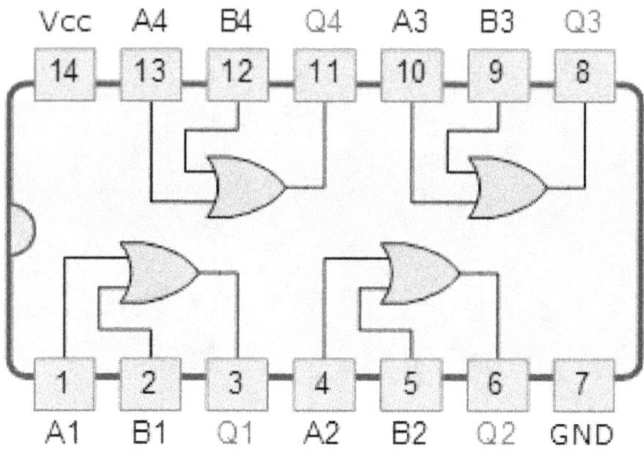

The Logic NOT Gate is the most basic of all the logical gates and is often referred to as an Inverting Buffer or simply an Inverter

Inverting NOT gates are single input devices which have an output level that is normally at logic level "1" and goes "LOW" to a logic level "0" when its single input is at logic level "1", in other words it "inverts" (complements) its input signal. The output from a NOT gate only returns "HIGH" again when its input is at logic level "0" giving us the Boolean expression of: A = Q.

Therefore, we can define the operation of a single input digital logic NOT gate as being:

"If A is NOT true, then Q is true"

Transistor Logic NOT Gate

A simple 2-input logic NOT gate can be constructed using an RTL Resistor-transistor switches as shown below with the input connected directly to the transistor base. The transistor must be saturated "ON" for an inverted output "OFF" at Q.

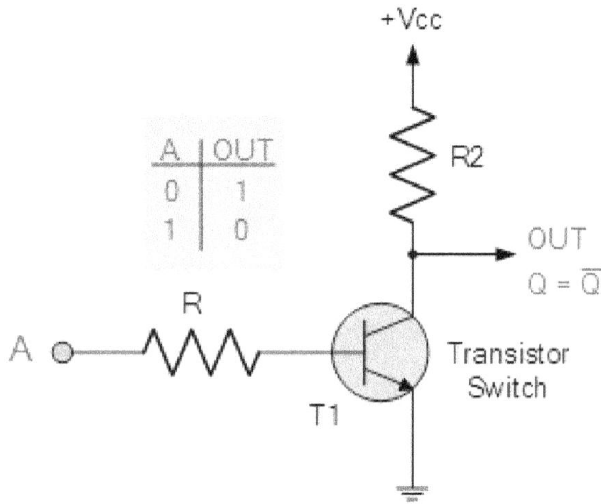

Logic NOT Gates are available using digital circuits to produce the desired logical function. The standard NOT gate is given a symbol whose shape is of a triangle pointing to the right with a circle at its end. This circle is known as an "inversion bubble" and is used in NOT, NAND and NOR symbols at their output to represent the logical operation of the NOT function. This bubble denotes a signal inversion (complementation) of the signal and can be present on either or both the output and/or the input terminals.

The Logic NOT Gate Truth Table

Symbol

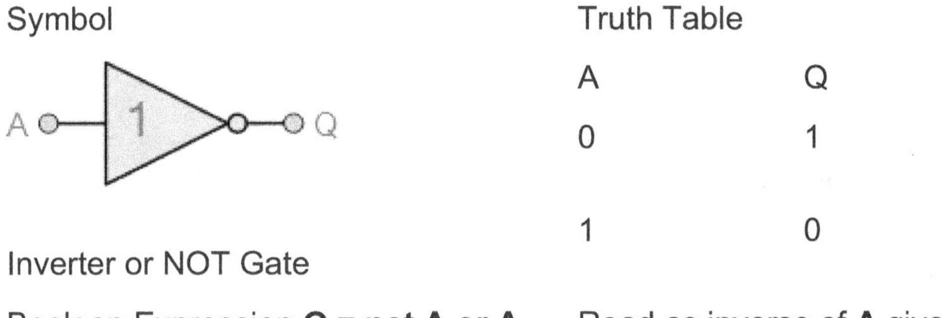

Truth Table

A	Q
0	1
1	0

Inverter or NOT Gate

Boolean Expression **Q = not A or A**   Read as inverse of **A** gives Q

Logic NOT gates provide the complement of their input signal. When their input signal is "HIGH" their output state will **NOT** be "HIGH". Similarly, when their input signal is "LOW" their output state will **NOT** be "LOW". Since they are single input devices, logic NOT gates are not normally classed as "decision" making devices or even as a gate, such as the AND or OR gates which have two or more logic inputs. Commercially available NOT gates ICs are available in either 4 or 6 individual gates within a single IC package.

The "bubble" (o) present at the end of the NOT gate symbol above denotes a signal inversion (complementation) of the output signal. But this bubble can Also be present at the gates input to indicate an *active-LOW* input. This inversion of the input signal is not restricted to the NOT gate only but can be used on any digital circuit or gate as shown with the operation of inversion being exactly the same whether on the input or output terminal. The easiest way is to think of the bubble as simply an inverter.

Signal Inversion using Active-low Input Bubble

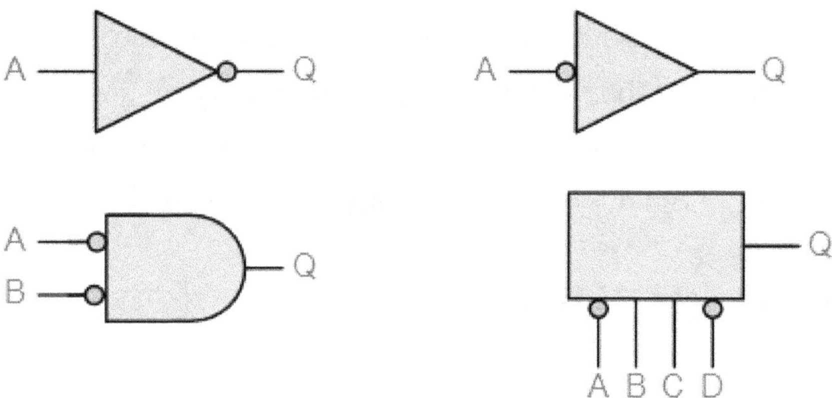

Bubble Notation for Input Inversion

NAND and NOR Gate Equivalents

An **Inverter** or logic NOT gate can Also be made using standard NAND and NOR gates by connecting together **ALL** their inputs to a common input signal for example.

A very simple inverter can Also be made using just a single stage transistor switching circuit as shown.

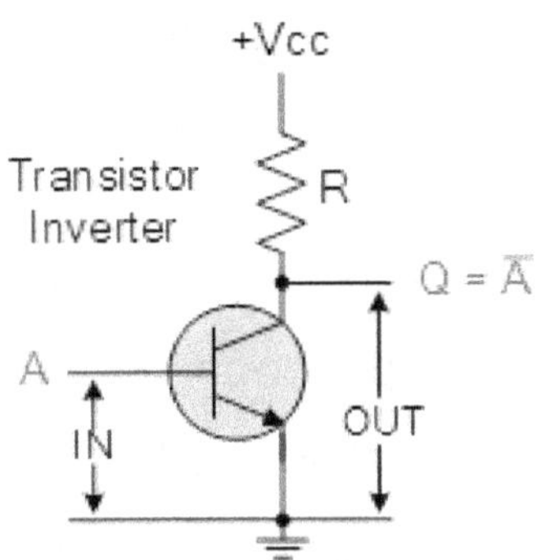

When the transistors base input at "A" is high, the transistor conducts and collector current flows producing a voltage drop across the resistor R thereby connecting the output point at "Q" to ground thus resulting in a zero voltage output at "Q".

Likewise, when the transistors base input at "A" is low (0v), the transistor now switches "OFF" and no collector current flows through the resistor resulting in an output voltage at "Q" high at a value near to +Vcc.

Then, with an input voltage at "A" HIGH, the output at "Q" will be LOW and an input voltage at "A" LOW the resulting output voltage at "Q" is HIGH producing the complement or inversion of the input signal.

Hex Schmitt Inverters

A standard **Inverter** or **Logic NOT Gate**, is usually made up from transistor switching circuits that do not switch from one state to the next instantly, there will always be Some delay in the switching action.

Also, as a transistor is a basic current amplifier, it can Also operate in a linear mode and any small variation to its input level will cause a variation to its output level or may even switch "ON" and "OFF" several times if there is any noise present in the circuit. One way to overcome these problems is to use a **Schmitt Inverter** or **Hex Inverter**.

We know that all digital gates use only two logic voltage states and that these are generally referred to as **Logic "1"** and **Logic "0"** any TTL voltage input between 2.0v and 5v is recognized as a logic "1" and any voltage input below 0.8v is recognized as a logic "0" respectively.

A **Schmitt Inverter** is designed to operate or switch state when its input signal goes above an "Upper Threshold Voltage" or **UTV** limit in which case the output changes and goes "LOW", and will remain in that state until the input signal falls below the "Lower Threshold Voltage" or **LTV** level in which case the output signal goes "HIGH". In other words, a Schmitt Inverter has Some form of **Hysteresis** built into its switching circuit.

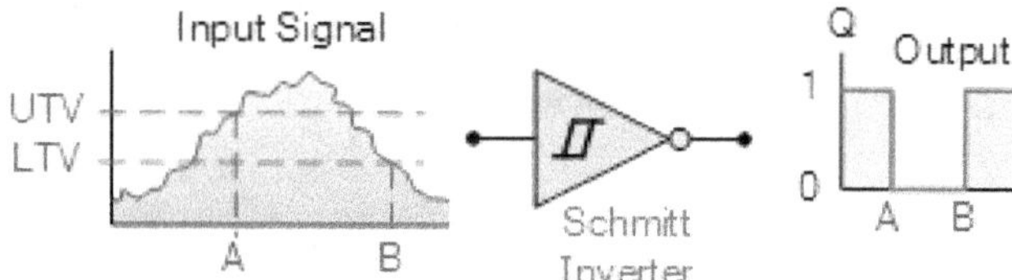

This switching action between an upper and lower threshold limit provides a much cleaner and faster "ON/OFF" switching output signal and makes the Schmitt inverter ideal for switching any slow-rising or slow-falling input signal and as such we can use a Schmitt trigger to convert these analogue signals into digital signals as shown.

Schmitt Inverter

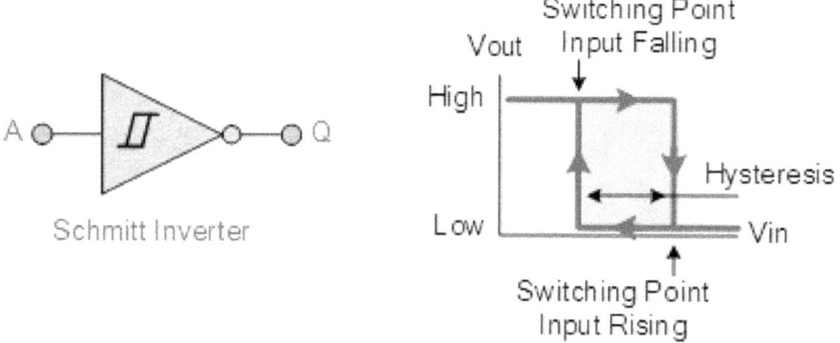

A very useful application of Schmitt inverters is when they are used as oscillators or sine-to-square wave converters for use as square wave clock signals.

Schmitt NOT Gate Inverter Oscillator

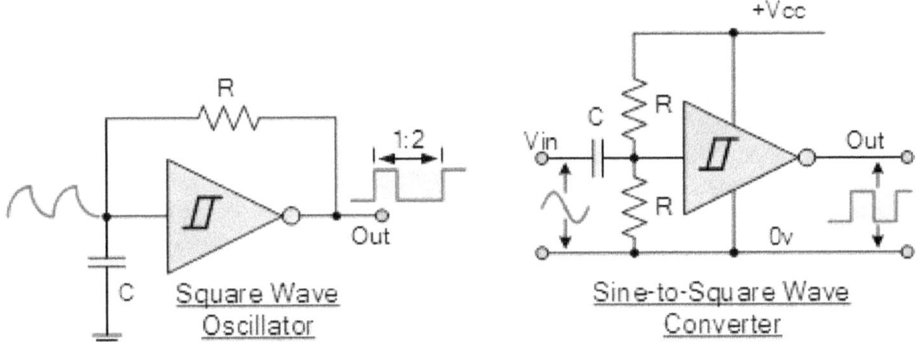

The first circuit shows a very simple low power RC type oscillator using a Schmitt inverter to generate a square wave output waveform. Initially the capacitor C is fully discharged Therefore the input to the inverter is "LOW" resulting in an inverted output which is "HIGH". As the output from the inverter is fed back to its input and the capacitor via the resistor R the capacitor begins to charge up.

When the capacitors charging voltage reaches the upper threshold limit of the inverter, the inverter changes state, the output becomes "LOW" and the capacitor begins to discharge through the resistor until it reaches the lower threshold level were the inverter changes state again. This switching back and forth by the inverter produces a square wave output signal with a 33% duty cycle and whose frequency is given as: $f = 680/RC$.

The second circuit converts a sine wave input (or any oscillating input for that matter) into a square wave output. The input to the inverter is connected to the junction of the potential divider network which is used to set the quiescent point of the circuit. The input capacitor blocks any DC component present in the input signal only allowing the sine wave signal to pass.

Since this signal passes the upper and lower threshold points of the inverter the output also changes from "HIGH" to "LOW" and therefore on producing a square wave output waveform. This circuit produces an output pulse on the positive rising edge of the input waveform, but by connecting a second Schmitt inverter to the output of the first, the basic circuit can be modified to produce an output pulse on the negative falling edge of the input signal.

Normally available logic NOT gate and Inverter ICs include:

<u>TTL Logic NOT Gates</u>

- 74LS04 Hex Inverting NOT Gate
- 74LS14 Hex Schmitt Inverting NOT Gate
- 74LS1004 Hex Inverting Drivers

<u>CMOS Logic NOT Gates</u>

- CD4009 Hex Inverting NOT Gate
- CD4069 Hex Inverting NOT Gate

7404 Logic NOT Gate or Inverter

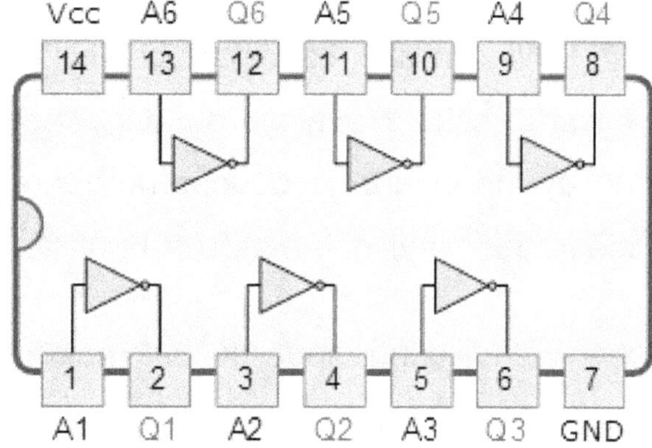

The Logic NAND Gate is a combination of a digital logic AND gate and a NOT gate connected together in series

The NAND (Not – AND) gate has an output that is normally at logic level "1" and only goes "LOW" to logic level "0" when **ALL** of its inputs are at logic level "1". The **Logic NAND Gate** is the reverse or "*Complementary*" form of the AND gate we have seen previously.

Logic NAND Gate Equivalence

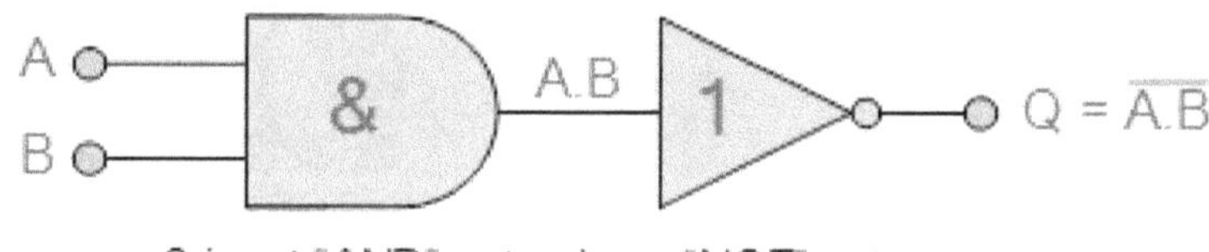

2-input "AND" gate plus a "NOT" gate

The logic or Boolean expression given for a logic NAND gate is that for *Logical Addition*, which is the opposite to the AND gate, and which it performs on the *complements* of the inputs. Its Boolean expression is denoted by a single dot or full stop symbol, (.) with a line or *Overline*, (‾) over the expression to signify the NOT or logical negation of the NAND gate giving us the Boolean expression of: A.B = Q.

Then we can define the operation of a 2-input digital device as being:

"If both A and B are true, then Q is NOT true"

Transistor NAND Gate

A simple 2-input NAND gate can be constructed using RTL Resistor-transistor switches connected together as shown below with the inputs connected directly to the transistor bases. Either transistor must be cut-off "OFF" for an output at Q.

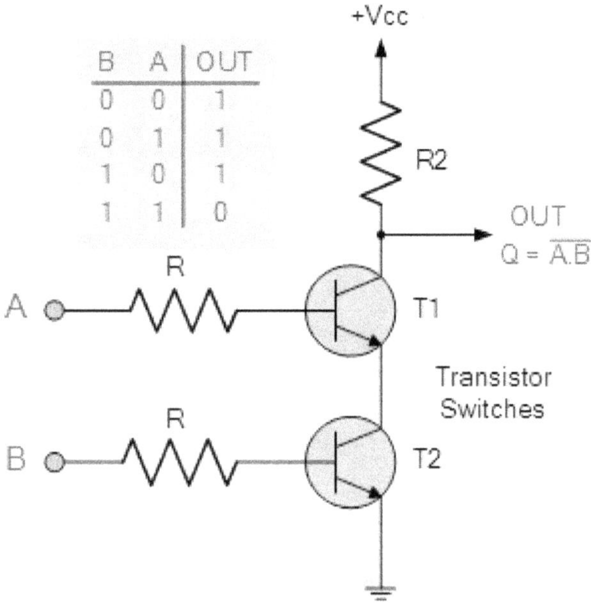

Logic NAND Gates are available using digital circuits to produce the desired logical function and is given a symbol whose shape is that of a standard AND gate with a circle, sometimes called an "inversion bubble" at its output to represent the NOT gate symbol with its logical operation given as:

The Digital Logic "NAND" Gate

2-input Function

Symbol

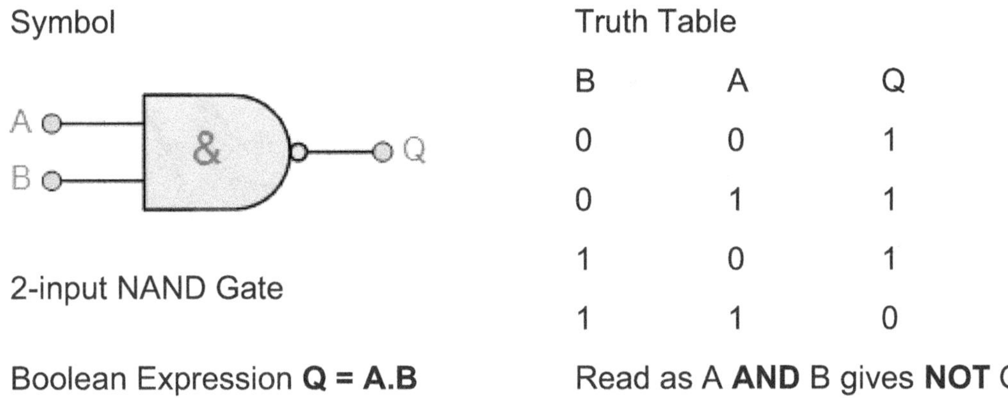

2-input NAND Gate

Boolean Expression **Q = A.B**

Truth Table

B	A	Q
0	0	1
0	1	1
1	0	1
1	1	0

Read as A **AND** B gives **NOT** Q

3-input Function

Symbol

3-input NAND Gate

Boolean Expression **Q = A.B.C**

Truth Table

C	B	A	Q
0	0	0	1
0	0	1	1
0	1	0	1
0	1	1	1
1	0	0	1
1	0	1	1
1	1	0	1
1	1	1	0

Read as A **AND** B **AND** C gives **NOT** Q

As with the AND function seen previously, the NAND function can Also have any number of individual inputs and commercially available NAND Gate ICs are available in standard 2, 3, or 4 input types. If additional inputs are required, then the standard NAND gates can be cascaded together to provide more inputs for example.

A 4-input Function

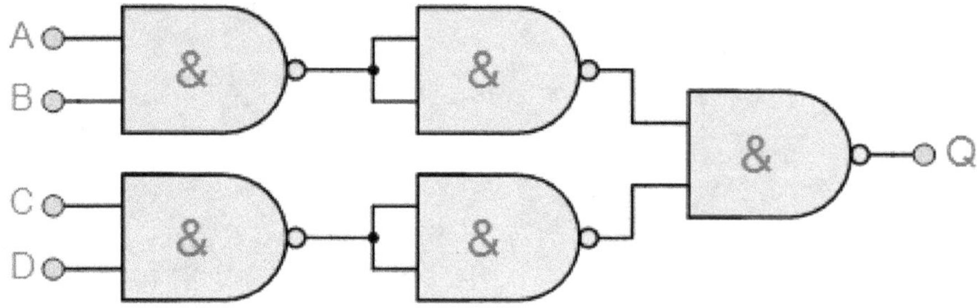

The Boolean Expression for this 4-input logic NAND gate will therefore be: **Q = A.B.C.D**

If the number of inputs required is an odd number of inputs any "unused" inputs can be held HIGH by connecting them directly to the power supply using suitable "Pull-up" resistors.

The **Logic NAND Gate** function is Sometimes known as the **Sheffer Stroke Function** and is denoted by a vertical bar or upwards arrow operator, for example, A NAND B = A|B or A↑B.

The "Universal" NAND Gate

The **Logic NAND Gate** is generally classed as a "Universal" gate as it is one of the most Normally used logic gate types. They can Also be used to produce any other type of logic gate function, but in practice being universal it forms the basis of most practical logic circuits.

By connecting them together in various combinations the three basic gate types of AND, OR and NOT function can be formed using only NAND gates, for example.

Various Logic Gates

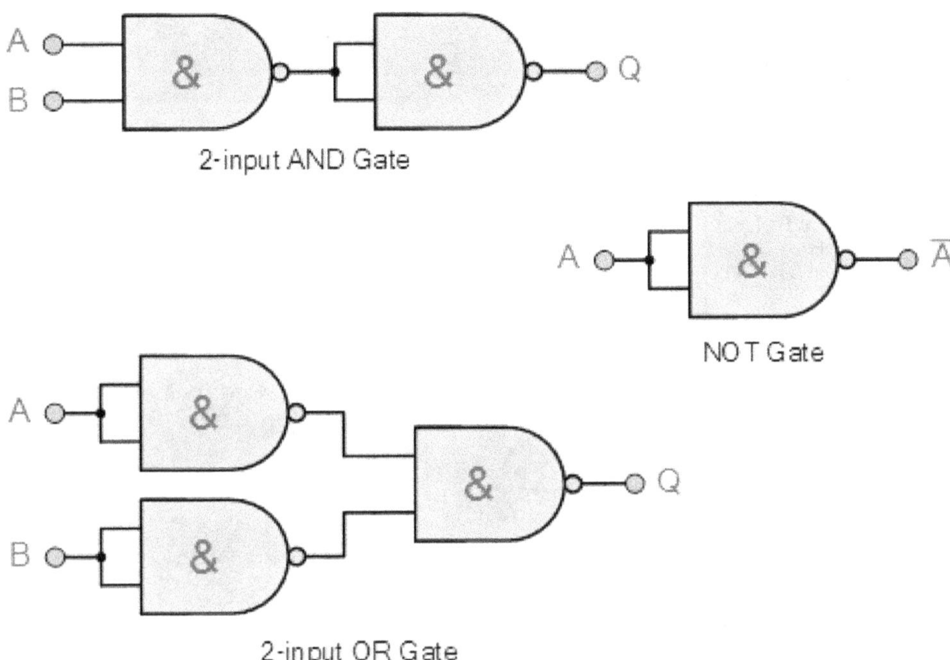

As well as the three common types above, Exclusive-OR, Exclusive-NOR and standard NOR gates can be formed using just individual NAND gates.

Normally available digital logic NAND gate ICs include:

<u>TTL Logic NAND Gates</u>

- 74LS00 Quad 2-input
- 74LS10 Triple 3-input
- 74LS20 Dual 4-input
- 74LS30 Single 8-input

<u>CMOS Logic NAND Gates</u>

- CD4011 Quad 2-input

- CD4023 Triple 3-input
- CD4012 Dual 4-input

7400 Quad 2-input Logic NAND Gate

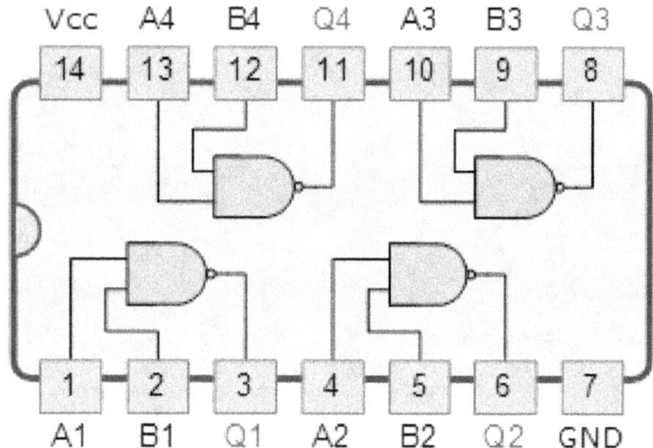

The Logic NOR Gate is a combination of the digital logic OR gate and an inverter or NOT gate connected together in series

The inclusive NOR (Not-OR) gate has an output that is normally at logic level "1" and only goes "LOW" to logic level "0" when **ANY** of its inputs are at logic level "1". The **Logic NOR Gate** is the reverse or "*Complementary*" form of the inclusive OR gate we have seen previously.

Logic NOR Gate Equivalent

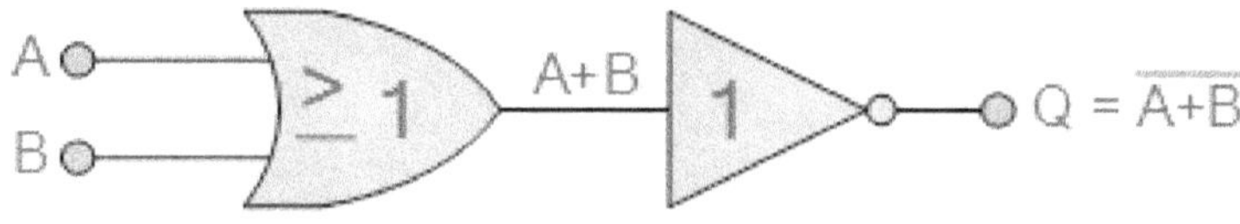

2-input "OR" gate plus a "NOT" gate

The logic or Boolean expression given for a logic NOR gate is that for *Logical Multiplication* which it performs on the *complements* of the inputs. Its Boolean expression is denoted by a plus sign, (+) with a line or *Overline*, (‾) over the expression to signify the NOT or logical negation of the NOR gate giving us the Boolean expression of: A+B = Q.

Then we can define the operation of a 2-input digital gate as being:

"If both A and B are NOT true, then Q is true"

Transistor Logic NOR Gate

A simple 2-input logic gate can be constructed using RTL Resistor-transistor switches connected together as shown below with the inputs connected directly to the transistor bases. Both transistors must be cut-off "OFF" for an output at Q.

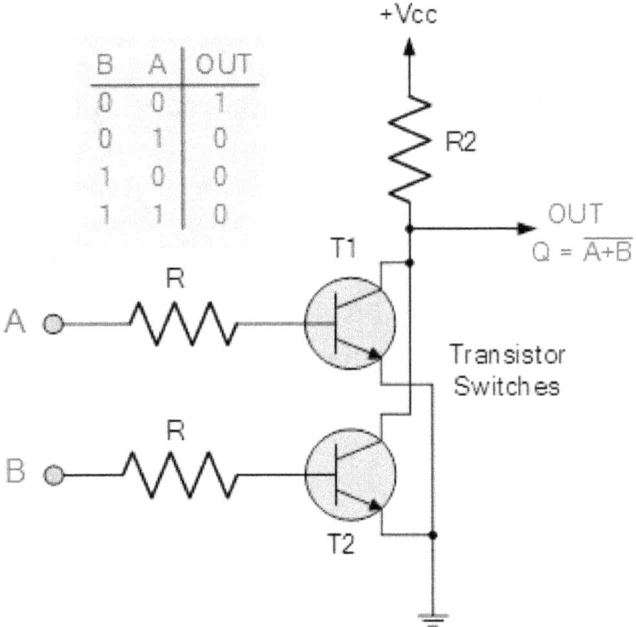

Logic NOR Gates are available using digital circuits to produce the desired logical function. It is given a symbol whose shape is that of a standard OR gate with a circle, sometimes called an "inversion bubble" at its output to represent the NOT (Inversion) gate symbol as shown:

The Digital Logic NOR Gate

2-input Logic NOR Gate

Symbol

2-input NOR Gate

Truth Table

B	A	Q
0	0	1
0	1	0
1	0	0
1	1	0

Boolean Expression **Q = A+B**   Read as A **OR** B gives **NOT** Q

3-input NOR Gate

Symbol

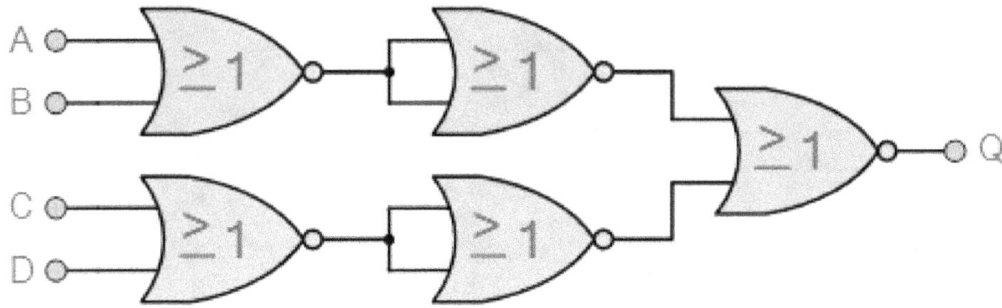

3-input NOR Gate

Truth Table

C	B	A	Q
0	0	0	1
0	0	1	0
0	1	0	0
0	1	1	0
1	0	0	0
1	0	1	0
1	1	0	0
1	1	1	0

Boolean Expression **Q = A+B+C**   Read as A **OR** B **OR** C gives **NOT** Q

As with the OR function, the NOR function can Also have any number of individual inputs and commercially available ICs are available in standard 2, 3, or 4 input types. If additional inputs are required, then the standard NOR gates can be cascaded together to provide more inputs for example.

A 4-input NOR Function

The Boolean Expression for this 4-input NOR gate will therefore be: **Q = A+B+C+D**

If the number of inputs required is an odd number of inputs any "unused" inputs can be held LOW by connecting them directly to ground using suitable "Pull-down" resistors.

The **Logic NOR Gate** function is Sometimes known as the **Pierce Function** and is denoted by a downwards arrow operator as shown, A↓B.

The "Universal" NOR Gate

NOR gate can Also be classed as a "Universal" type gate. NOR gates can be used to produce any other type of logic gate function just like the NAND gate and by connecting them together in various combinations the three basic gate types of AND, OR and NOT function can be formed using only NOR gates, for example.

Various Logic Gates using only NOR Gates

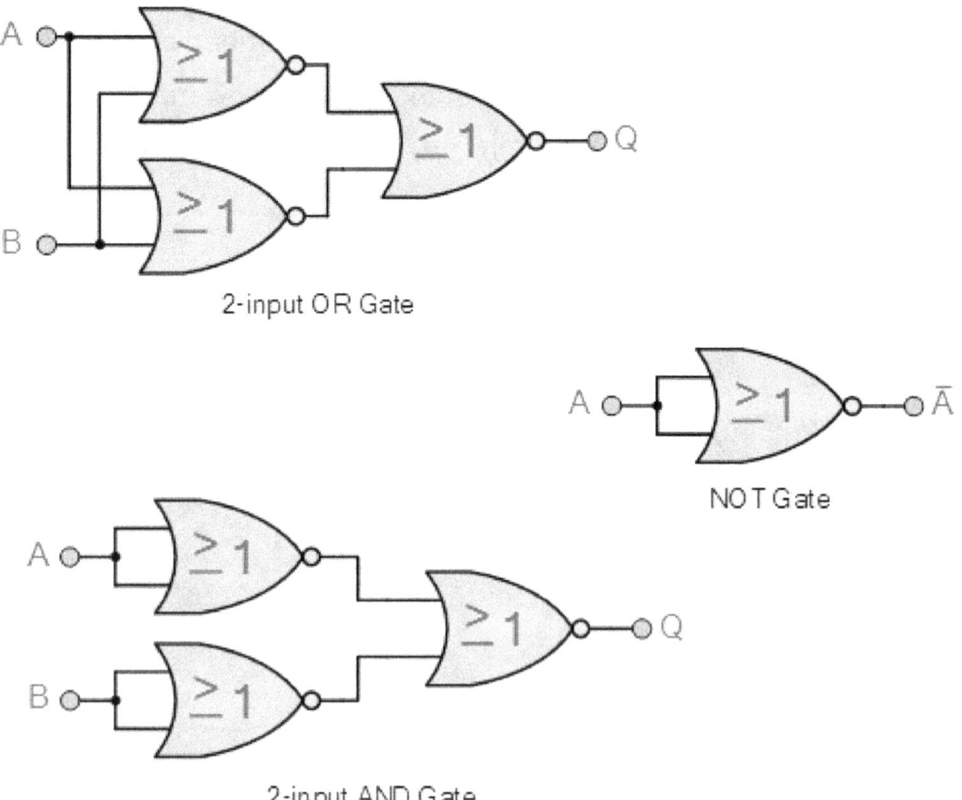

Same like the three common types above, Exclusive-OR, Exclusive-NOR and standard NOR gates can also be formed using just individual NOR gates.

Normally available digital logic NOR gate ICs include:

<u>TTL Logic NOR Gates</u>

- 74LS02 Quad 2-input
- 74LS27 Triple 3-input
- 74LS260 Dual 4-input

<u>CMOS Logic NOR Gates</u>

- CD4001 Quad 2-input
- CD4025 Triple 3-input
- CD4002 Dual 4-input

7402 Quad 2-input Logic NOR Gate

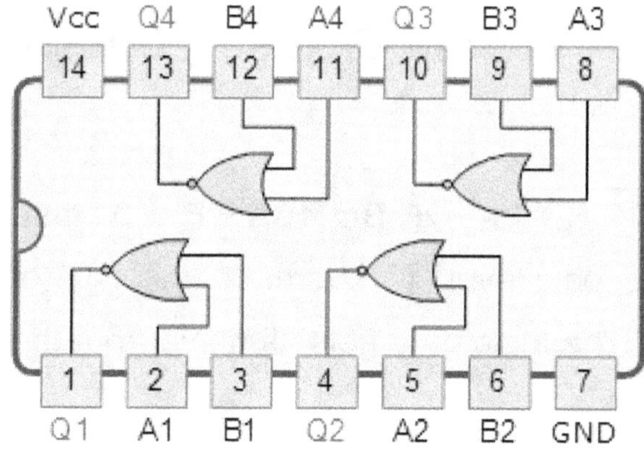

The Exclusive-OR logic function is a very useful circuit that can be used in many different types of computational circuits

The "Exclusive OR Gate" is another type of digital logic gate Normally used in arithmetic operations since it can be used to give the sum of two binary numbers as well as error-detection and correction circuits.

It has been observed that by using the three principal gates, the AND Gate, the OR Gate and the NOT Gate, we can build many other types of logic gate functions, such as a NAND Gate and a NOR Gate or any other type of digital logic function, we can imagine.

But there are two other types of digital logic gates which although they are not a basic gate in their own right as they are constructed by combining together other logic gates, their output Boolean function is important enough to be considered as complete logic gates. These two "hybrid" logic gates are called the **Exclusive-OR (Ex-OR) Gate** and its complement the **Exclusive-NOR (Ex-NOR) Gate**.

We know that for a 2-input OR gate, if A = "1", **OR** B = "1", **OR BOTH** A + B = "1" then the output from the digital gate must Also be at a logic level "1". Therefore, this type of logic gate is known as an Inclusive-OR function. The logic gate gets its name from the fact that it *includes* the case of Q = "1" when both A and B = "1".

If a logic output "1" is obtained when **ONLY** A = "1" or when **ONLY** B = "1" but **NOT** both together at the same time, giving the binary inputs of "01" or "10", then the output will be "1". This type of gate is known as an Exclusive-OR function or more Normally an Ex-Or function for short. Since Boolean expression *excludes* the "**OR BOTH**" case of Q = "1" when both A and B = "1", this thing happens.

In other words, the output of an Exclusive-OR gate **ONLY** goes "HIGH" when its two input terminals are at "**DIFFERENT**" logic levels with respect to each other.

An odd number of logics "1's" on its inputs gives a logic "1" at the output. These two inputs can be at logic level "1" or at logic level "0" giving us the Boolean expression of: $Q = (A \oplus B) = A.B + A.B$

The **Exclusive-OR Gate** function, or **Ex-OR** for short, is achieved by combining standard logic gates together to form more complex gate functions that are used extensively in building arithmetic logic circuits, computational logic comparators and error detection circuits.

The two-input "Exclusive-OR" gate is basically a modulo two adders, since it gives the sum of two binary numbers and as a result are more complex in design than other basic types of logic gate. The truth table, logic symbol and implementation of a 2-input Exclusive-OR gate is shown below.

The Digital Logic "Exclusive-OR" Gate

2-input Ex-OR Gate

Symbol

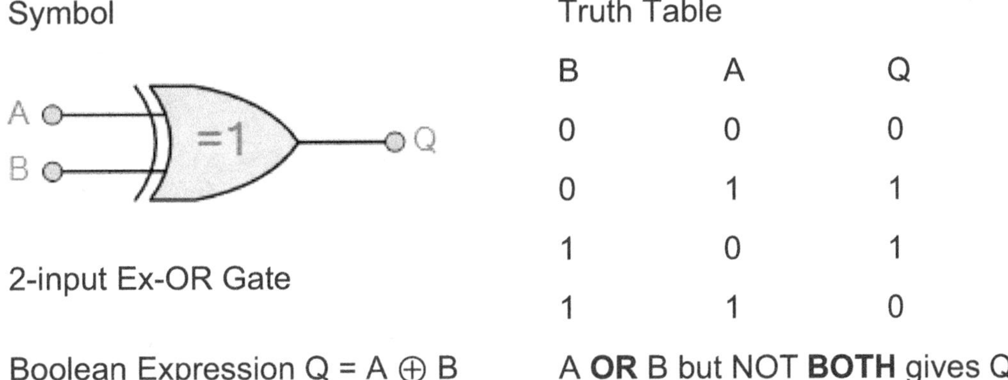

2-input Ex-OR Gate

Boolean Expression $Q = A \oplus B$

Truth Table

B	A	Q
0	0	0
0	1	1
1	0	1
1	1	0

A **OR** B but NOT **BOTH** gives Q

Giving the Boolean expression of: $Q = AB + AB$

The truth table above shows that the output of an Exclusive-OR gate ONLY goes "HIGH" when both of its two input terminals are at "DIFFERENT" logic levels with respect to each other. If these two inputs, A and B are both at logic level "1" or both at logic level "0" the output is a "0" making the gate an "odd but not the even gate". In other words, the output is "1" when there are an odd number of 1's in the inputs.

This ability of the *Exclusive-OR gate* to compare two logic levels and produce an output value dependent upon the input condition is very useful in computational logic circuits as it gives us the following Boolean expression of:

$Q = (A \oplus B) = A.B + A.B$

The logic function implemented by a 2-input Ex-OR is given as either: "A OR B but NOT both" will give an output at Q. In general, an Ex-OR gate will give an output value of logic "1" ONLY when there are an **ODD** number of 1's on the inputs to the gate, if the two numbers are equal, the output is "0".

Then an Ex-OR function with more than two inputs is called an "odd function" or modulo-2-sum (Mod-2-SUM), not an Ex-OR. This description can be expanded to apply to any number of individual inputs as shown below for a 3-input Ex-OR gate.

3-input Ex-OR Gate

Symbol

3-input Ex-OR Gate

Truth Table

C	B	A	Q
0	0	0	0
0	0	1	1
0	1	0	1
0	1	1	0
1	0	0	1
1	0	1	0
1	1	0	0

Boolean Expression $Q = A \oplus B \oplus C$ "Any **ODD** Number of Inputs" gives Q

Giving the Boolean expression of: Q = ABC + ABC + ABC + ABC

The symbol used to denote an Exclusive-OR odd function is slightly different to that for the standard Inclusive-OR Gate. The logic or Boolean expression given for a logic OR gate is that of logical addition which is denoted by a standard plus sign.

The symbol used to describe the Boolean expression for an **Exclusive-OR** function is a plus sign, (+) within a circle (O). This exclusive-OR symbol Also represents the mathematical "direct sum of sub-objects" expression, with the resulting symbol for an *Exclusive-OR* function being given as: ($\oplus$).

We said previously that the Ex-OR function is not a basic logic gate but a combination of different logic gates connected together. Using the 2-input truth table above, we can expand the Ex-OR function to: (A+B).(A.B) which means that we can understand this new expression using the following individual gates.

Ex-OR Gate Equivalent Circuit

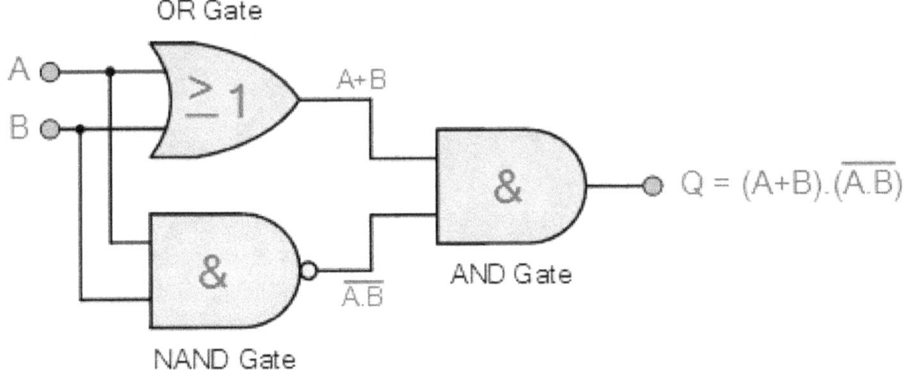

One of the main disadvantages of implementing the Ex-OR function above is that it contains three different types logic gates OR, NAND and finally AND within its design. One easier way of producing the Ex-OR function from a single gate is to use our old favorite the NAND gate as shown below.

Understanding Ex-OR Function using NAND gates

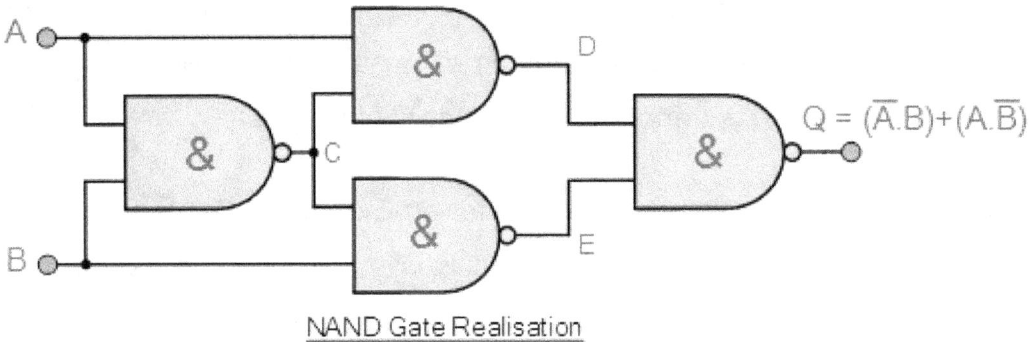

NAND Gate Realisation

Exclusive-OR Gates are used mainly to build circuits that perform arithmetic operations and calculations especially **Adders** and **Half-Adders** as they can provide a "carry-bit" function or as a controlled inverter, where one input passes the binary data and the other input is supplied with a control signal.

Normally available digital logic Exclusive-OR gate ICs include:

TTL Logic Ex-OR Gates

- 74LS86 Quad 2-input

CMOS Logic Ex-OR Gates

- CD4030 Quad 2-input

7486 Quad 2-input Exclusive-OR Gate

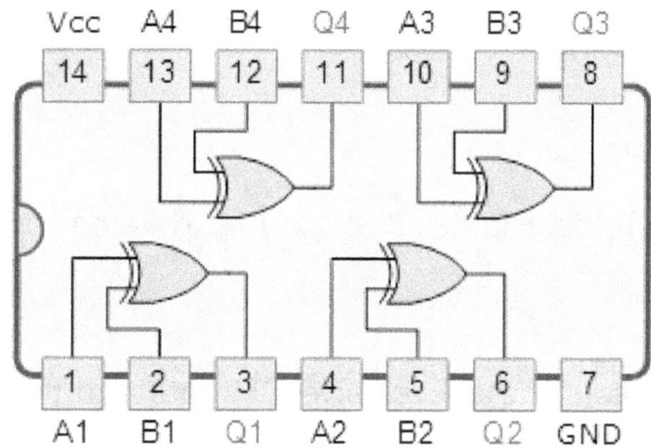

The **Exclusive-OR** logic function is a very useful circuit that can be used in many different types of computational circuits. Although not a basic logic gate in its own right, its usefulness and versatility has turned it into a standard logical function complete with its own Boolean expression, operator and symbol. The *Exclusive-OR Gate* is widely available as a standard quad two-input 74LS86 TTL gate or the 4030B CMOS package.

One of its most Normally used applications is as a basic logic comparator which produces a logic "1" output when its two input bits are not equal. As of this, the exclusive-OR gate has an inequality status being known as an odd function. In order to compare numbers that contain two or more bits, additional exclusive-OR gates are needed with the 74LS85 logic comparator being 4-bits wide.

The Exclusive-NOR Gate function is a digital logic gate that is the reverse or complementary form of the Exclusive-OR function

Basically the "Exclusive-NOR Gate" is a combination of the Exclusive-OR gate and the NOT gate but has a truth table similar to the standard NOR gate in that it has an output that is normally at logic level "1" and goes "LOW" to logic level "0" when **ANY** of its inputs are at logic level "1".

However, an output "1" is only obtained if **BOTH** of its inputs are at the same logic level, either binary "1" or "0". For example, "00" or "11". This input combination would then give us the Boolean expression of: $Q = (A \oplus B) = A.B + A.B$

Then the output of a digital logic Exclusive-NOR gate **ONLY** goes "HIGH" when its two input terminals, A and B are at the "**SAME**" logic level which can be either at a logic level "1" or at a logic level "0". In other words, an even number of logic "1's" on its inputs gives a logic "1" at the output, otherwise is at logic level "0".

Then this type of gate gives and output "1" when its inputs are *"logically equal"* or *"equivalent"* to each other, which is why an **Exclusive-NOR** gate is Sometimes called an **Equivalence Gate**.

The logic symbol for an Exclusive-NOR gate is simply an Exclusive-OR gate with a circle or "inversion bubble", (o) at its output to represent the NOT function. Then the **Logic Exclusive-NOR Gate** is the reverse or *"Complementary"* form of the Exclusive-OR gate, $(A \oplus B)$ we have seen previously.

Ex-NOR Gate Equivalent

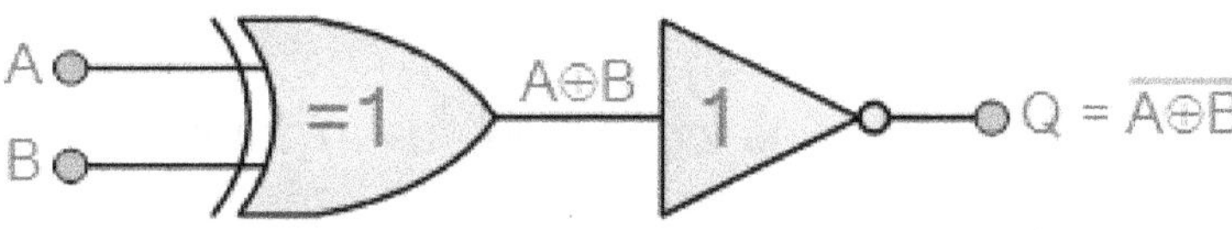

2-input "Ex-OR" gate plus a "NOT" gate

The **Exclusive-NOR Gate**, also written as: "Ex-NOR" or "XNOR", function is achieved by combining standard gates together to form more complex gate functions and an example of a 2-input Exclusive-NOR gate is given below.

The Digital Logic "Exclusive NOR" Gate

2-input Exclusive NOR Gate

Symbol

2-input Ex-NOR Gate

Boolean Expression $Q = A \oplus B$

Truth Table

B	A	Q
0	0	1
0	1	0
1	0	0
1	1	1

Read if A **AND** B the **SAME** gives Q

Giving the Boolean expression of: Q = AB + AB

The logic function implemented by a 2-input Ex-NOR gate is given as "when both A AND B are the SAME" will give an output at Q. In general, an Exclusive-NOR gate will give an output value of logic "1" ONLY when there are an **EVEN** number of 1's on the inputs to the gate (the inverse of the Ex-OR gate) except when all its inputs are "LOW".

Then an Ex-NOR function with more than two inputs is called an "even function" or modulo-2-sum (Mod-2-SUM), not an Ex-NOR. This description can be expanded to apply to any number of individual inputs as shown below for a 3-input Exclusive-NOR gate.

3-input Exclusive NOR Gate

Symbol

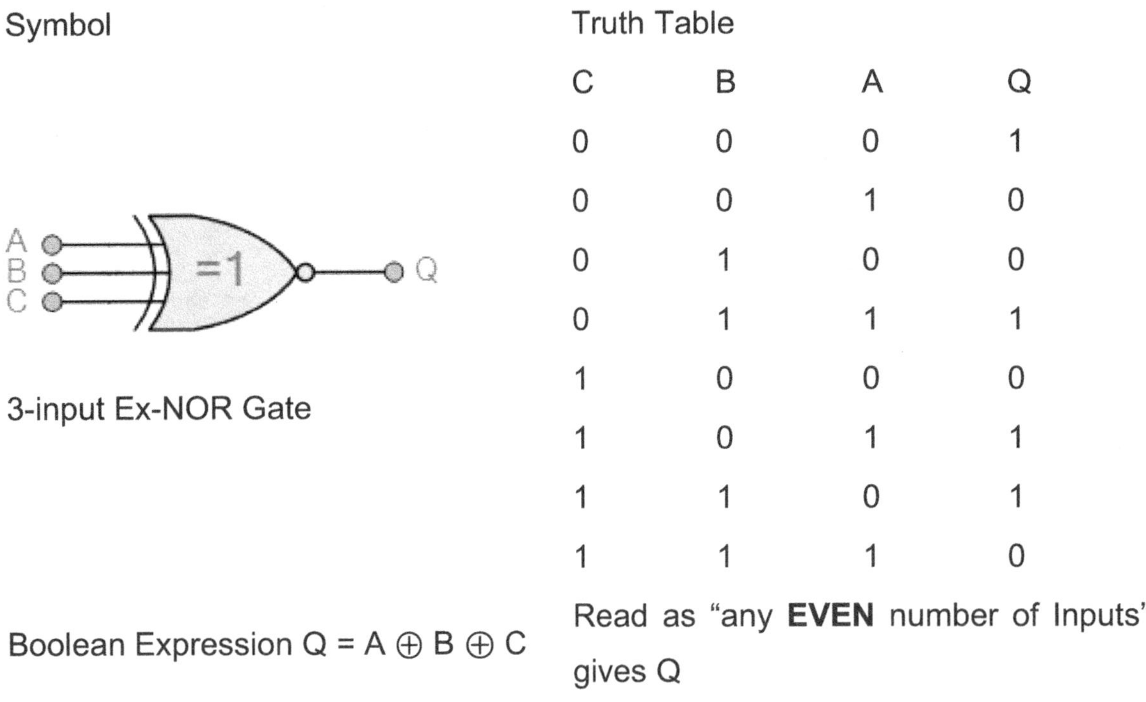

3-input Ex-NOR Gate

Boolean Expression Q = A ⊕ B ⊕ C

Truth Table

C	B	A	Q
0	0	0	1
0	0	1	0
0	1	0	0
0	1	1	1
1	0	0	0
1	0	1	1
1	1	0	1
1	1	1	0

Read as "any **EVEN** number of Inputs" gives Q

Giving the Boolean expression of: Q = ABC + ABC + ABC + ABC

We said previously that the Ex-NOR function is a combination of different basic logic gates Ex-OR and a NOT gate, and by using the 2-input truth table above, we can expand the Ex-NOR function to: Q = A ⊕ B = (A.B) + (A.B) which means we can understand this new expression using the following individual gates.

Ex-NOR Gate Equivalent Circuit

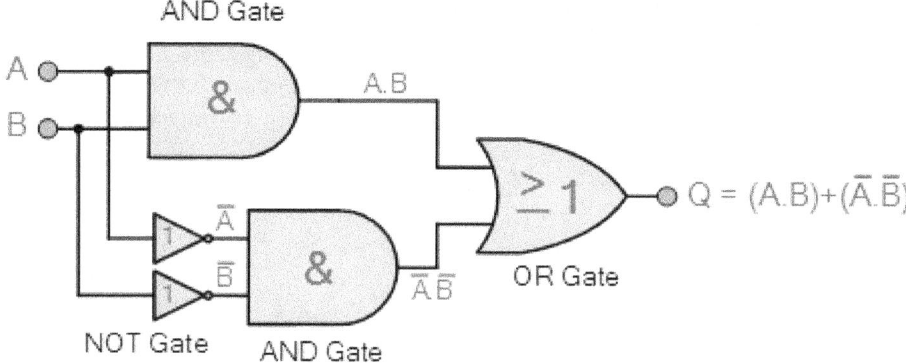

One of the main disadvantages of implementing the Ex-NOR function above is that it contains three different types logic gates the AND, NOT and finally an OR gate within its basic design. One easier way of producing the Ex-NOR function from a single gate type is to use NAND gates as shown below.

Understanding Ex-NOR Function using NAND gates

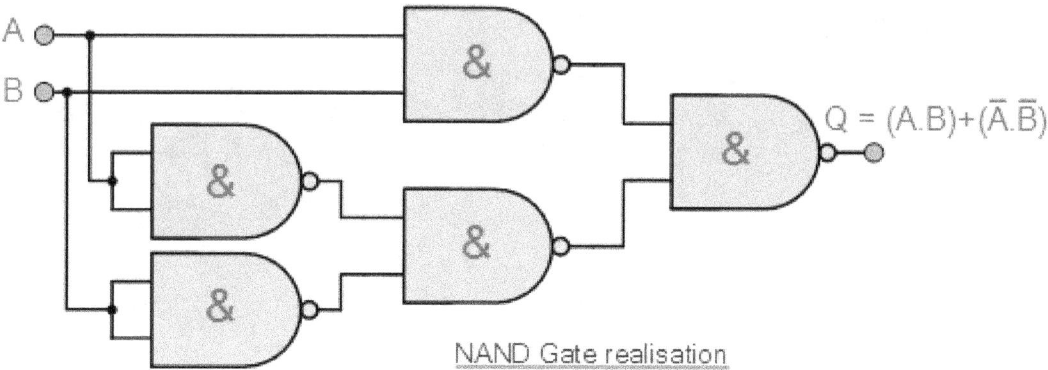

Ex-NOR gates are used mainly in electronic circuits that perform arithmetic operations and data checking such as *Adders*, *Subtractors* or *Parity Checkers*, etc. As the Ex-NOR gate gives an output of logic level "1" whenever its two inputs are equal it can be used to compare the magnitude of two binary digits or numbers and Therefore Ex-NOR gates are used in Digital Comparator circuits.

Normally available digital logic Exclusive-NOR gate ICs include:

<u>TTL Logic Ex-NOR Gates</u>

- 74LS266 Quad 2-input

<u>CMOS Logic Ex-NOR Gates</u>

- CD4077 Quad 2-input

74266 Quad 2-input Ex-NOR Gate

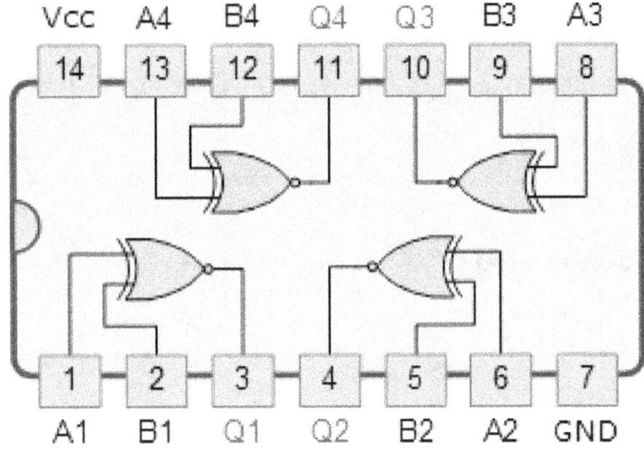

Digital Buffers and Tri-state Buffers can provide current amplification in a digital circuit to drive output loads

The digital buffer is the logic gate opposite of an inverter (Not Gate) we look at in the previous chapter where we saw that the NOT gates output state is the complement, opposite or inverse of its input signal.

Therefore, for example, when the single input to NOT gate is "HIGH", its output state will NOT be "HIGH". When its input signal is "LOW" its output state will NOT be "LOW", in other words it "inverts" its input signal, hence the name "Inverter".

In digital electronic circuits, often it is required to isolate logic gates from each other or have them drive or switch higher than normal loads, like relays, solenoids and lamps without the need for inversion. One type of single input logic gate that allows us to do just that is known as the **Digital Buffer**.

Unlike the single input, single output inverter or NOT gate such as the TTL 7404 which inverts or complements its input signal on the output, the "Buffer" performs no inversion or decision-making capabilities (like logic gates with two or more inputs) but instead produces an output which exactly matches that of its input. In other words, a digital buffer does nothing as its output state equals its input state.

Then digital buffers can be regarded as Idempotent gates applying Boole's Idempotent Law As when an input passes through this device its value is not changed. Therefore, the digital buffer is a "non-inverting" device and will therefore give us the Boolean expression of: $Q = A$.

Then we can define the logical operation of a single input digital buffer as being:

"Q is true, only when A is true"

In other words, the output (Q) state of a buffer is only true (logic "1") when its input A is true, otherwise its output is false (logic "0").

The Single Input Digital Buffer

Symbol   Truth Table

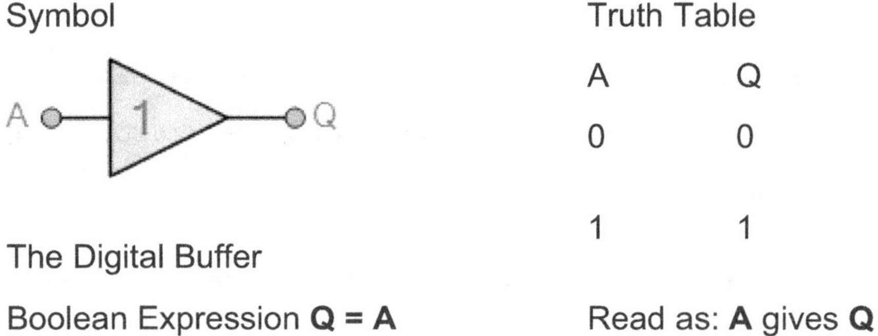

The Digital Buffer

Boolean Expression **Q = A**   Read as: **A** gives **Q**

The **Digital Buffer** can also be made by connecting together two NOT gates as shown below. The first will "invert" the input signal A and the second will "re-invert" it back to its original level performing a double inversion of the input.

Double Inversion using NOT Gates

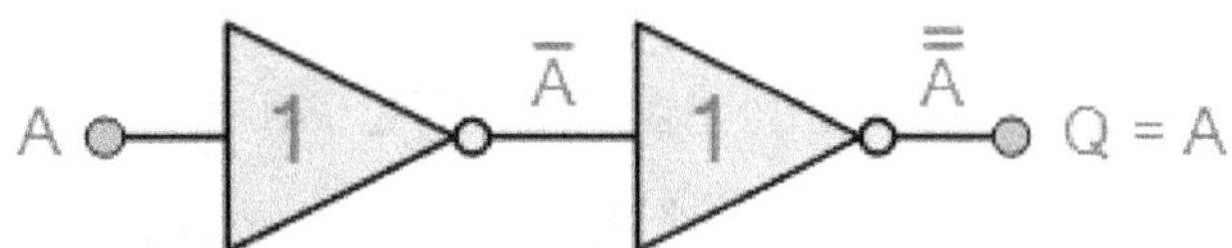

You may be thinking, well what's the point of a Digital Buffer if it does not invert or alter its input signal in any way, or make any logical decisions or operations like the AND or OR gates do, then why not just use a piece of wire instead, and that's a good point. But a non-inverting Digital Buffer does have many uses in digital electronics with one of its main advantages being that it provides digital amplification.

Digital Buffers can be used to isolate other gates or circuit stages from each other preventing the impedance of one circuit from affecting the impedance of another. A digital buffer can Also be used to drive high current loads such as transistor switches

Since their output drive capability is generally much higher than their input signal requirements. In other words, buffers can be used for power amplification of a digital signal as they have what is called a high "fan-out" capability.

Digital Buffer Fan-out Example

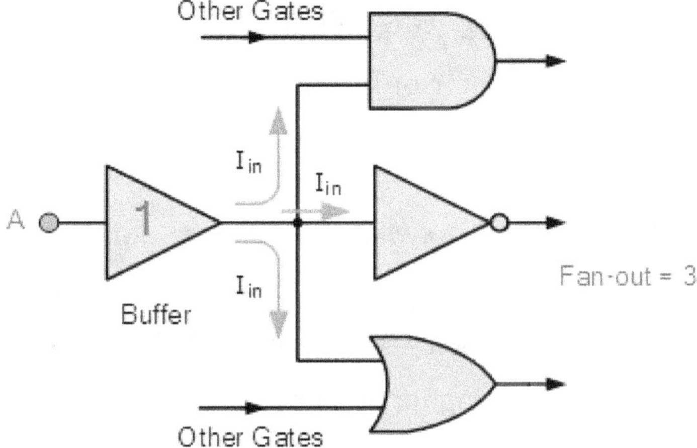

The **Fan-out** parameter of a buffer (or any digital IC) is the output driving capability or output current capability of a logic gate giving greater power amplification of the input signal. It may be necessary to connect more than just one logic gate to the output of another or to switch a high current load such as an LED, then a Buffer will allow us to do just that.

The output of a logic gate is usually connected to the inputs of other gates. Each input requires a certain amount of current from the gate output to change state, therefore that each additional gate connection adds to the load of the gate. Therefore, the fan-out is the number of parallel loads that can be driven simultaneously by one digital buffer of logic gate. Acting as a current Source a buffer can have a high fan-out rating of up to 20 gates of the same logic family.

If a digital buffer has a high fan-out rating (current source) it must Also have a high "fan-in" rating (current sink) as well. However, the propagation delay of the gate deteriorates rapidly as a function of fan-in. Therefore, gates with a fan-in greater than 4 should be avoided.

Applications where it is required to decouple gates from each other, a **Tri-state Buffer** or tristate output driver can be used.

The "Tri-state Buffer"

As well as the standard Digital Buffer seen above, there is another type of digital buffer circuit whose output can be "electronically" disconnected from its output circuitry when required. This type of Buffer is known as a 3-State Buffer or more Normally a Tri-state Buffer.

A Tri-state Buffer can be thought of as an input controlled switch with an output that can be electronically turned "ON" or "OFF" by means of an external "Control" or "Enable" (EN) signal input. This control signal can be either a logic "0" or a logic "1" type signal resulting in the Tri-state Buffer being in one state allowing its output to operate normally producing the required output or in another state where its output is blocked or disconnected.

Then a tri-state buffer requires two inputs. One being the data input and the other being the enable or control input as shown.

Tri-state Buffer Switch Equivalent

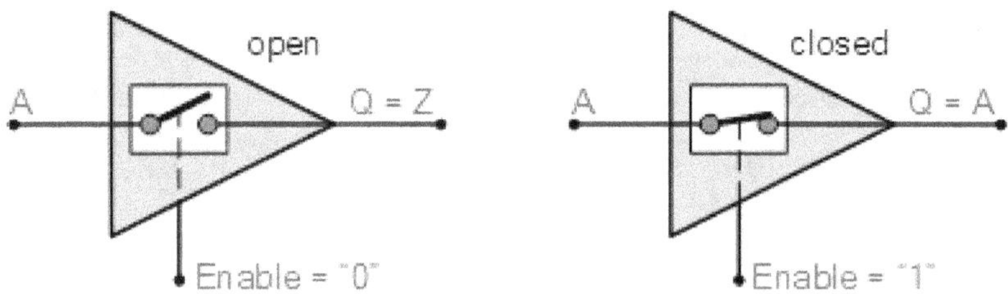

When activated into its third state it disables or turns "OFF" its output producing an open circuit condition that is neither at a logic "HIGH" or "LOW", but instead gives an output state of very high impedance, **High-Z**, or more Normally Hi-Z. Then this type of device has two logic state inputs, "0" or a "1" but can produce three different output states, "0", "1" or " Hi-Z " which is why it is called a "Tri" or "3-state" device.

Note that this third state is NOT equal to a logic level "0" or "1", but is a high impedance state in which the buffers output is electrically disconnected from the rest of the circuit. As a result, no current is drawn from the supply.

There are four different types of Tri-state Buffer, one set whose output is enabled or disabled by an **"Active-HIGH"** control signal producing an inverted or non-inverted output, and another set whose buffer output is controlled by an **"Active-LOW"** control signal producing an inverted or non-inverted output as shown below.

Active "HIGH" Tri-state Buffer

Symbol

Truth Table

Enable	IN	OUT
0	0	Hi-Z
0	1	Hi-Z
1	0	0
1	1	1

Tri-state Buffer

Read as Output = Input if Enable is equal to "1"

An **Active-high** Tri-state Buffer such as the 74LS241 octal buffer, is activated when a logic level "1" is applied to its "enable" control line and the data passes through from its input to its output. When the enable control line is at logic level "0", the buffer output is disabled and a high impedance condition, Hi-Z is present on the output.

An active-high tri-state buffer can Also have an inverting output as well as its high impedance state creating an active-high tri-state inverting buffer as shown.

Active "HIGH" Inverting Tri-state Buffer

Symbol

Truth Table

Enable	IN	OUT
0	0	Hi-Z

0	1	Hi-Z
1	0	1
1	1	0

Read as Output = Inverted Input if Enable equals "1"

The output of an active-high inverting tri-state buffer, such as the 74LS240 octal buffer, is activated when a logic level "1" is applied to its "enable" control line. The data at the input is passes through to the output but is inverted producing a complement of the input. When the enable line is LOW at logic level "0", the buffer output is disabled and at a high impedance condition, Hi-Z.

The same two tri-state buffers can Also be implemented with an active-low enable input as shown.

Active "LOW" Tri-state Buffer

Symbol

Tri-state Buffer

Truth Table

Enable	IN	OUT
0	0	0
0	1	1
1	0	Hi-Z
1	1	Hi-Z

Read as Output = Input if Enable is **NOT** equal to "1"

An **Active-low** Tri-state Buffer is the opposite to the above, and is activated when a logic level "0" is applied to its "**enable**" control line. The data passes through from its input to its output. When the enable control line is at logic level "1", the buffer output is disabled and a high impedance condition, Hi-Z is present on the output.

Active "LOW" Inverting Tri-state Buffer

Symbol

Truth Table

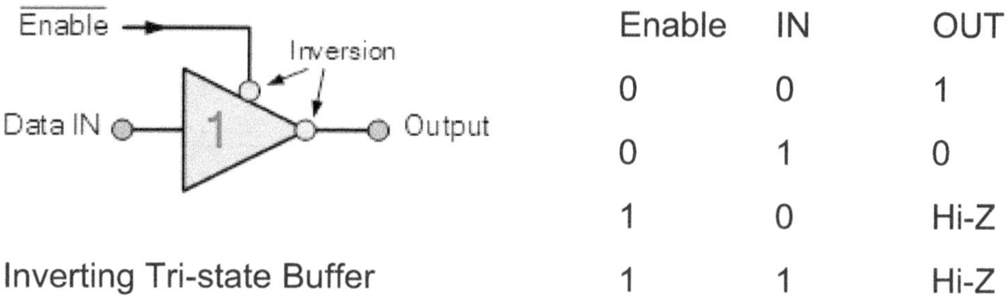

Enable	IN	OUT
0	0	1
0	1	0
1	0	Hi-Z
1	1	Hi-Z

Inverting Tri-state Buffer

Read as Output = Inverted Input if Enable is **NOT** equal to "1"

An **Active-low** Inverting Tri-state Buffer is the opposite to the above as its output is enabled or disabled when a logic level "0" is applied to its "**enable**" control line. When a buffer is enabled by a logic "0", the output is the complement of its input. When the enable control line is at logic level "1", the buffer output is disabled and a high impedance condition, Hi-Z is present on the output.

Tri-state Buffer Control

We have seen above that a buffer can provide voltage or current amplification within a digital circuit and it can Also be used to invert the input signal. We have Also seen that digital buffers are available in the tri-state form that allows the output to be effectively switched-off producing a high impedance state (Hi-Z) equivalent to an open circuit.

The Tri-state Buffer is used in many electronic and microprocessor circuits as they allow multiple logic devices to be connected to the same wire or bus without damage or loss of data. For example, suppose we have a data line or data bus with Some memory, peripherals, I/O or a CPU connected to it. Each of these devices is capable of sending or receiving data to each other onto this single data bus at the same time creating what is called a contention.

Contention occurs when multiple devices are connected together As Some want to drive their output high and some low. If these devices start to send or receive data at the same time a short circuit may occur when one device outputs to the bus a logic "1", the supply voltage, while another is set at logic level "0" or ground, resulting in a short circuit condition and possibly damage to the devices as well as loss of data.

Digital information is sent over these data buses or data highways either serially, one bit at a time, or it may be up to eight (or more) wires together in a parallel form such as in a microprocessor data bus allowing multiple tri-state buffers to be connected to the same data highway without damage or loss of data as shown.

Tri-state Digital Buffer Data Bus Control

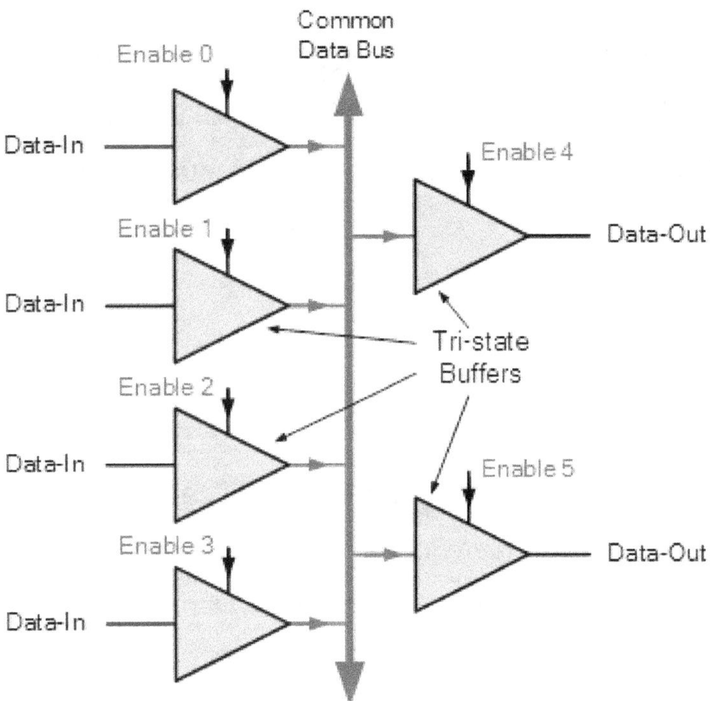

Then, the **Tri-state Buffer** can be used to isolate devices and circuits from the data bus and one another. If the outputs of several Tri-state Buffers are electrically connected together Decoders are used to allow only one set of Tri-state Buffers to be active at any one time while the other devices are in their high impedance state. An example of Tri-state Buffers connected to a 4-wire data bus is shown below.

Tri-state Digital Buffer Control

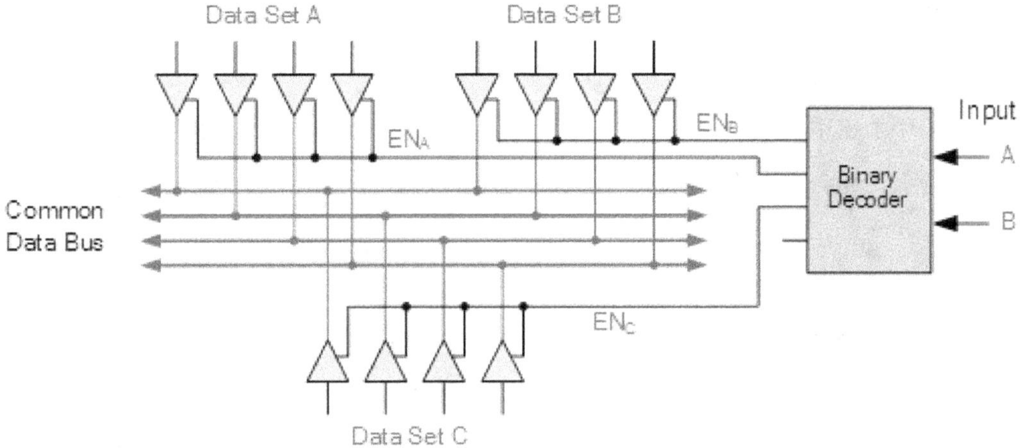

This basic example shows how a binary decoder can be used to control a number of tri-state buffers either individually or together in data sets. The decoder selects the appropriate output that corresponds to its binary input allowing only one set of data to pass either a logic "1" or logic "0" output state onto the bus. At this time all the other tri-state outputs connected to the same bus lines are disabled by being placed in their high impedance Hi-Z state.

Then data from data set "A" can only be transferred to the common bus when an active HIGH signal is applied to the tri-state buffers via the Enable line, EN_A. At all other times it represents a high impedance condition effectively being isolated from the data bus.

Likewise, data set "B" only passes data to the bus when an enable signal is applied via EN_B. A good example of tri-state buffers connected together to control data sets is the TTL 74244 Octal Buffer.

It is Also possible to connect Tri-state Buffers "back-to-back" to produce what is called a **Bi-directional Buffer** circuit with one "active-high buffer" connected in parallel but in reverse with one "active-low buffer".

Here, the "enable" control input acts more like a directional control signal causing the data to be both read "from" and transmitted "to" the same data bus wire. In this type of application, a tri-state buffer with bi-directional switching capability such as the TTL 74245 can be used.

We have seen that a **Tri-state buffer** is a non-inverting device which gives an output (which is same as its input) only when the input to the Enable, (EN) pin is HIGH otherwise the output of the buffer goes into its high impedance, (Hi-Z) state. Tri-state outputs are used in many integrated circuits and digital systems and not just in digital tristate buffers.

Both digital buffers and tri-state buffers can be used to provide voltage or current amplification driving much high loads such as relays, lamps or power transistors than with conventional logic gates. But a buffer can Also be used to provide electrical isolation between two or more circuits.

We have seen that a data bus can be created if several tristate devices are connected together and as long as only one is selected at any one time, there is no problem. Tri-state buses allow several digital devices to input and output data on the same data bus by using I/O signals and address decoding.

Tri-state Buffers are available in integrated form as quad, hex or octal buffer/drivers in both uni-directional and bi-directional forms, with the more common being the TTL 74240, the TTL 74244 and the TTL 74245 as shown.

Normally available **Digital Buffer** and **Tri-state Buffer** ICs include:

<u>TTL Logic Digital Buffers</u>

- 74LS07 Hex Non-inverting Buffer
- 74LS17 Hex Buffer/Driver
- 74LS244 Octal Buffer/Line Driver
- 74LS245 Octal Bi-directional Buffer

<u>CMOS Logic Digital Buffers</u>

- CD4050 Hex Non-inverting Buffer
- CD4503 Hex Tri-state Buffer
- HEF40244 Tri-state Octal Buffer

74LS07 Digital Buffer

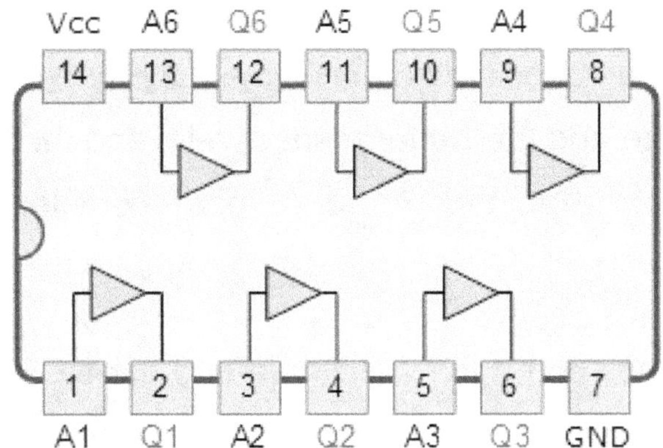

74LS244 Octal Tri-state Buffer

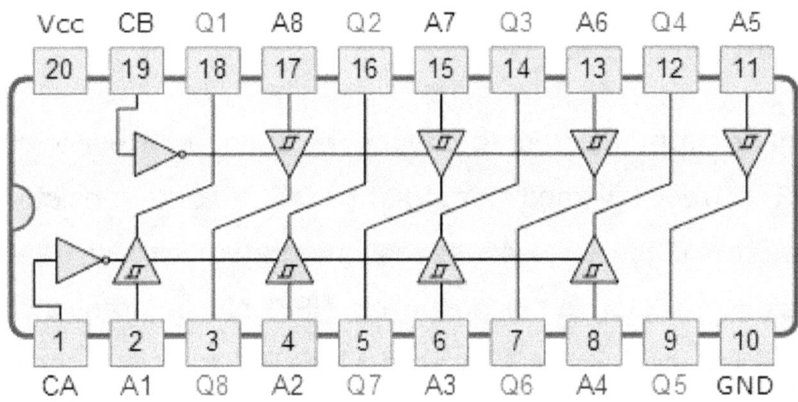

There are three basic types of digital logic gates, the AND Gate, the OR Gate and the NOT Gate.

It has been observed that digital logic gates have an opposite or complementary form of itself in the form of the NAND Gate, the NOR Gate and the Buffer respectively, and that any of these individual gates can be connected together to form more complex Combinational Logic circuits.

It has Also been observed that in digital electronics both the NAND gate and the NOR gate can both be classed as "**Universal**" gates as they can be used to construct any other gate type. In fact, any combinational circuit can be constructed using only two or three input NAND or NOR gates. We Also saw that NOT gates and Buffers are single input devices that can Also have a **Tri-state** High-impedance output which can be used to control the flow of data onto a common data bus wire.

Digital Logic Gates can be made from discrete components such as Resistors, Transistors and Diodes to form **RTL** (resistor-transistor logic) or **DTL** (diode-transistor logic) circuits, but today's modern digital 74xxx series integrated circuits are manufactured using **TTL** (transistor-transistor logic) based on NPN bipolar transistor technology or the much faster and low power CMOS based MOSFET transistor logic used in the 74Cxxx, 74HCxxx, 74ACxxx and the 4000 series logic chips.

The eight most "standard" individual **Digital Logic Gates** are summarized below along with their corresponding truth tables.

Standard Logic Gates

The Logic AND Gate

Symbol

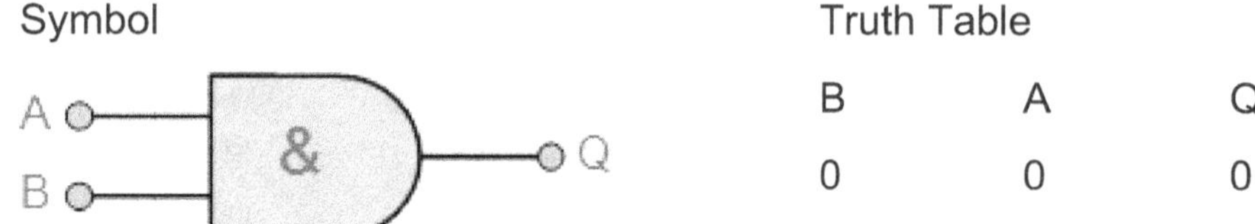

Truth Table

B	A	Q
0	0	0

0	1	0
1	0	0
1	1	1

Boolean Expression **Q = A.B**

Read as A **AND** B gives Q

The Logic OR Gate

Symbol

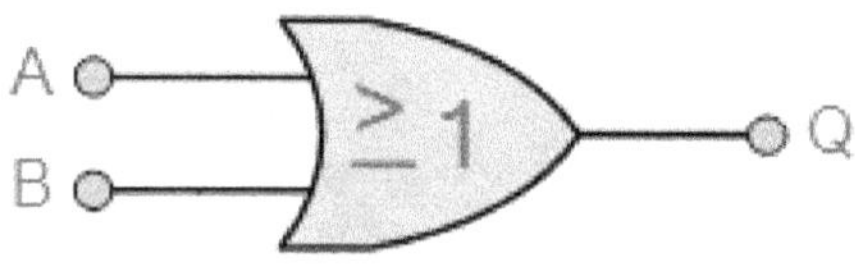

Truth Table

B	A	Q
0	0	0
0	1	1
1	0	1
1	1	1

Boolean Expression Q = A + B

Read as A **OR** B gives Q

Inverting Logic Gates

The Logic NAND Gate

Symbol

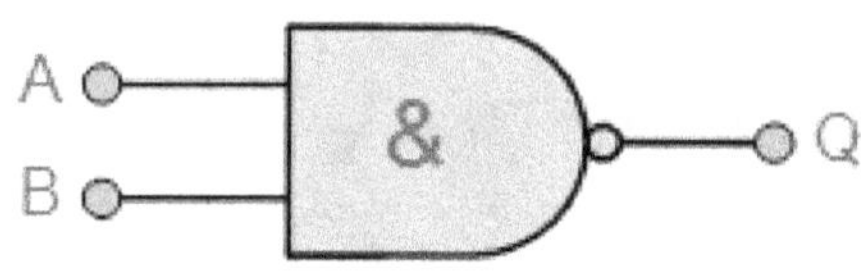

Truth Table

B	A	Q
0	0	1
0	1	1
1	0	1
1	1	0

Boolean Expression Q = A . B

Read as A **AND** B gives **NOT** Q

The Logic NOR Gate

Symbol

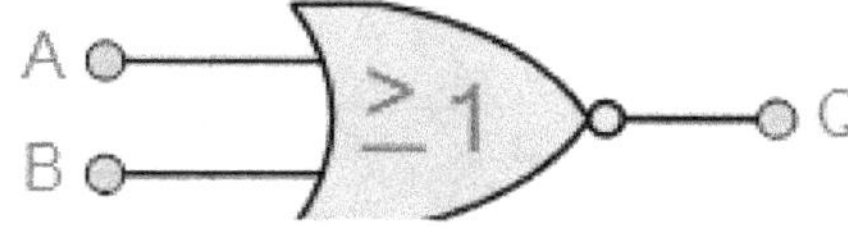

Truth Table

B	A	Q
0	0	1

0	1	0
1	0	0
1	1	0

Boolean Expression Q = A + B   Read as A **OR** B gives **NOT** Q

Exclusive Logic Gates

The Logic Exclusive-OR Gate (Ex-OR)

Symbol   Truth Table

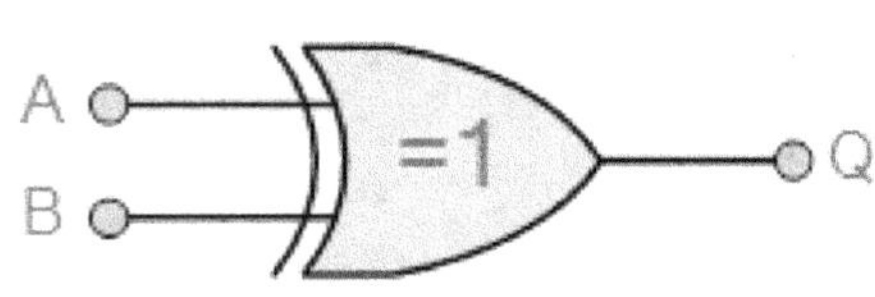

B	A	Q
0	0	0
0	1	1
1	0	1
1	1	0

Boolean Expression $Q = A \oplus B$ Read as A **OR** B but not **BOTH** gives Q (odd)

The Logic Exclusive-NOR Gate (Ex-NOR)

Symbol   Truth Table

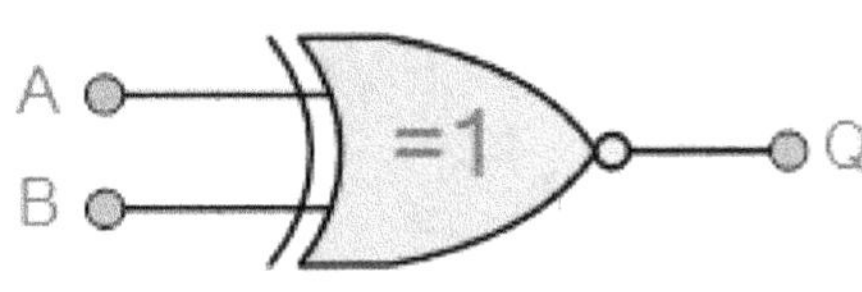

B	A	Q
0	0	1
0	1	0
1	0	0
1	1	1

Boolean Expression $Q = A \oplus B$ Read if A **AND** B the **SAME** gives Q (even)

Single Input Logic Gates

The Hex Buffer

Symbol   Truth Table

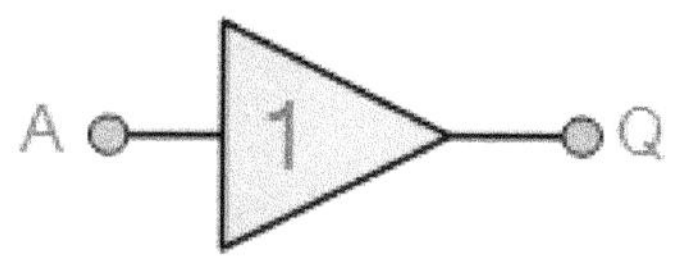

A	Q
0	0
1	1

Boolean Expression Q = A

Read as **A** gives **Q**

The NOT gate (Inverter)

Symbol

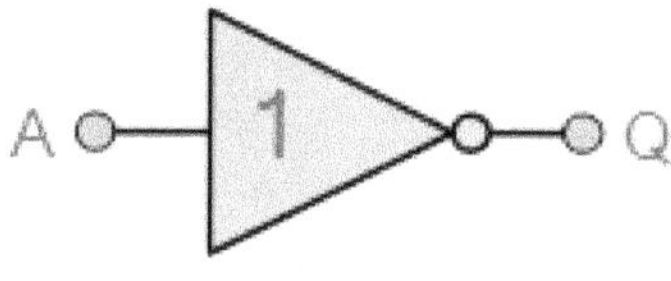

Boolean Expression Q = not A or A

Truth Table

A	Q
0	1
1	0

Read as inverse of **A** gives Q

The operation of the above **Digital Logic Gates** and their Boolean expressions can be summarized into a single truth table as shown below. This truth table shows the relationship between each output of the main digital logic gates for each possible input combination.

Digital Logic Gate Truth Table Summary

The following logic gates truth table compares the logical functions of the 2-input logic gates detailed above.

Inputs		Truth Table Outputs for Each Gate					
B	A	AND	NAND	OR	NOR	EX-OR	EX-NOR
0	0	0	1	0	1	0	1
0	1	0	1	1	0	1	0
1	0	0	1	1	0	1	0
1	1	1	0	1	0	0	1

Truth Table Output for Single-input Gates

A	NOT	Buffer
0	1	0
1	0	1

Pull-up and Pull-down Resistors

One final point to remember, when connecting together digital logic gates to produce logic circuits, any "unused" inputs to the gates must be connected directly to either a logic level "1" or a logic level "0" by means of a suitable "Pull-up" or "Pull-down" resistor (for example 1kΩ resistor) to produce a fixed logic signal. This will prevent the unused input to the gate from "floating" about and producing false switching of the gate and circuit.

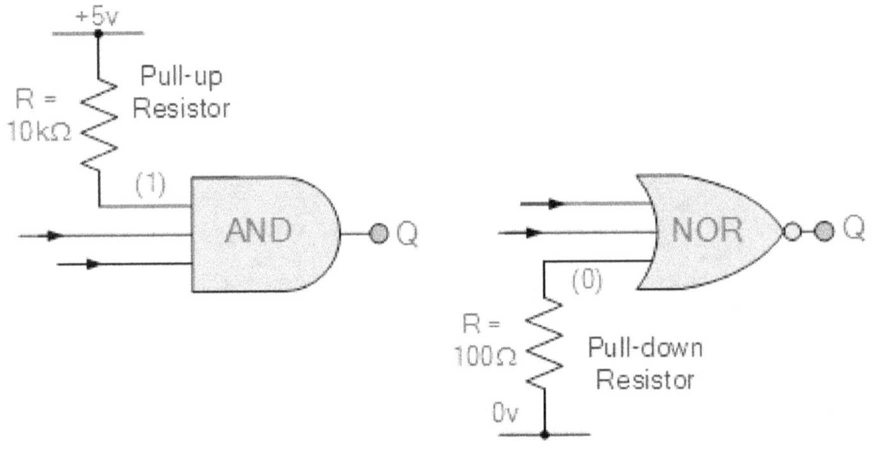

As well as using pull-up or pull-down resistors to prevent unused logic gates from floating about, spare inputs to gates and latches can Also be connected together or connected to left-over or spare gates within a single IC package as shown.

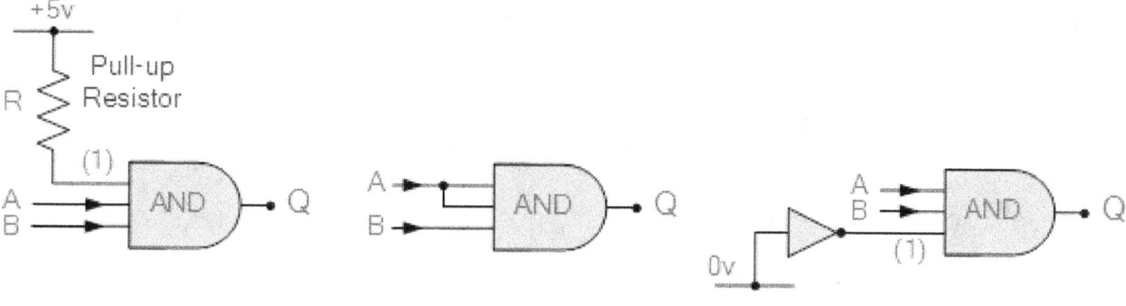

Pull-up and Pull-down resistors are used to correctly bias the inputs of digital gates to stop them from floating about randomly when there is no input condition

Digital logic gates can be used for connection to external circuits or devices but care must be taken to ensure that their inputs or outputs function correctly and provide the expected switching condition, and **Pull-up Resistors** do just that.

Modern digital logic gates, ICs and micro-controllers contain many inputs, called "pins" as well as one or more outputs, and these inputs and outputs need to be correctly set, either HIGH or LOW for the digital circuit to function correctly.

We know that logic gates are the most basic building block of any digital logic circuit and that by using combinations of the three basic gates, the AND gate, the OR gate and NOT gate, we can construct quite complex combinational circuits. But being digital, these circuits can only have one of two logic states, called the logic "0" state or the logic "1" state.

These logic states are represented by two different voltage levels with any voltage below one level regarded as a logic "0", and any voltage above another level regarded as logic "1". Therefore, for example, if the two voltage levels are 0V and +5V, then the 0V represents a logic "0" and the +5V represents a logic "1".

If the inputs to a digital logic gate or circuit are not within the range by which it can be sensed as either a logic "0" or a logic "1" input, then the digital circuit may false trigger as the gate or circuit does not recognize the correct input value, as the HIGH may not be high enough or the LOW may not be low enough.

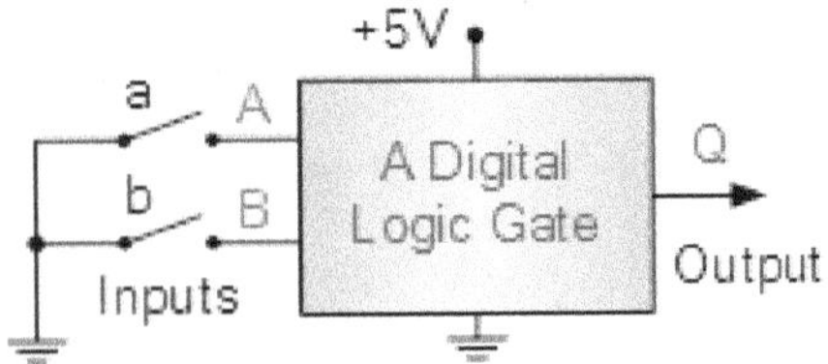

For example, consider the digital circuit on the left. The two switches, "a" and "b", represent the inputs to a generic logic gate. When switch "a" is closed (ON), input "A" is connected to ground, (0v) or logic level "0" (LOW) and likewise, when switch "b" is closed (ON), input "B" is Also connected to ground, logic level "0" (LOW) and this is the correct condition we require.

However, when switch "a" is opened (OFF), what will be the value of the voltage applied to input "A", HIGH or LOW? We assume it will be +5V (HIGH) as switch "a" is open-circuited and therefore input "A" is not shorted to ground, but this may not be the case. As the input is now effectively unconnected from either a defined HIGH or LOW condition, it has the potential to "float" about between 0V and +5V (Vcc) allowing the input to self–bias at any voltage level whether that represents a HIGH or a LOW condition.

This uncertain situation may cause the digital input at "A" to stay at a logic level "0" (LOW) when the switch is open, when we actually need a logic "1", (HIGH) causing the logic gate to falsely switch the output at "Q". Also once there, this floating and weak input signal could easily change value at the slightest of interference or noise from its neighboring inputs or could even cause it to go into oscillation, rendering the gate practically unusable. The same situation is Also true with regards to the switching of input "B".

Then to prevent accidental switching of digital circuits, any unconnected inputs called "floating inputs" should be tied to a logic "1" or logic "0" as appropriate for the circuit. We can easily do this by using what are Normally called **Pull-up Resistors** and **Pull-down Resistors** to give the input pin a defined default state, even if the switch is open, closed or there is nothing is connected to it.

When building digital electronic circuits, generally you will have Some spare gates or latches within a single IC package left over, or the design of the circuit results in not all of a multi-input gates inputs being used. These unused logic inputs can be tied together or connected to a fixed voltage, using a high value resistor to either the Vcc voltage, known as pull-up or via a low value resistor to 0V (GND), known as pull-down. These unused inputs should never be left just floating about.

Pull-up Resistors

The most common method of ensuring that the inputs of digital logic gates and circuits cannot self-bias and float about is to either connect the unused pins directly to ground (0V) for a constant low "0" input, (OR and NOR gates) or directly to Vcc (+5V) for a constant high "1" input (AND and NAND gates). Ok, let's look again at our two switched inputs from above.

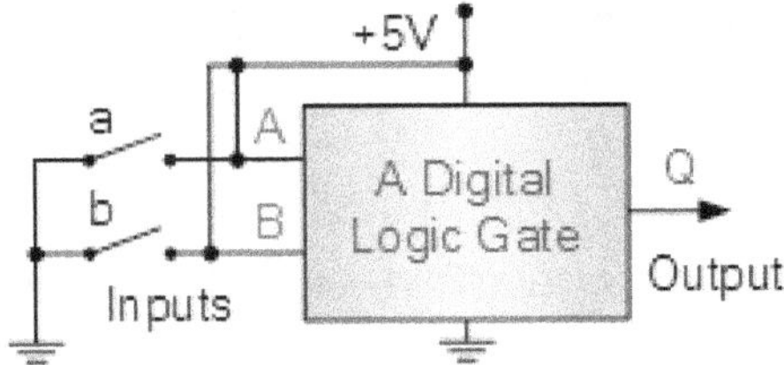

This time, to stop the two inputs, A and B, from "floating" about when the corresponding switches, "a" and "b" are open (OFF), the two inputs are connected to +5V supply.

You may think that this would work fine as when switch "a" is open (OFF), the input is connected to Vcc (+5V) and when the switch is closed (ON), the input is connected to ground as before, then inputs "A" or "B" always have a default state regardless of the position of the switch.

However, this is a bad condition as when either of the switches are closed (ON), there will be a direct short circuit between the +5V supply and ground, resulting in excessive current flow either blowing a fuse or damaging the circuit which is not good news. One way to overcome this issue is to use a pull-up resistor connected between the input pin and the +5V supply rail as shown.

Pull-up Resistor Application

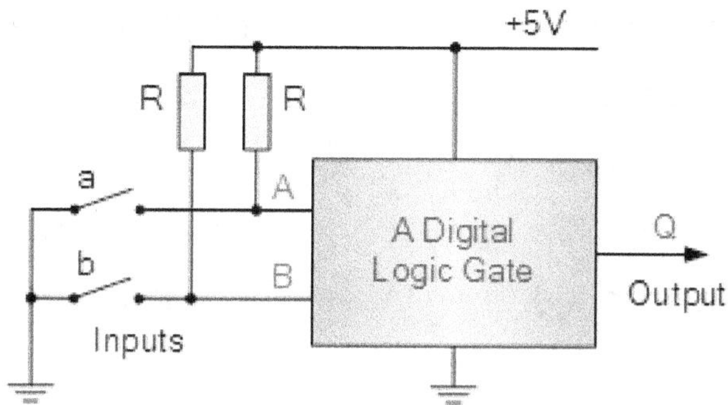

By using these two pull-up resistors, one for each input, when switch "A" or "B" is open (OFF), the input is effectively connected to the +5V supply rail via the pull-up resistor. Since there is very little input current into the input of the logic gate, very little voltage is dropped across the pull-up resistor. As a result, nearly all the +5V supply voltage is applied to the input pin creating a HIGH, logic "1" condition.

When switches "A", or "B" are closed, (ON) the input is shorted to ground (LOW) creating a logic "0" condition as before at the input. However, this time we are not shorting out the supply rail as the pull-up resistor only passes a small current (as determined by Ohms law) through the closed switch to ground.

By using a *pull-up resistor* in this way, the input always has a default logic state, either "1" or "0", high or low, depending on the position of the switch, thus achieving the proper output function of the gate at "Q" and therefore preventing the input from floating about or self-biasing giving us exactly the switching condition we require.

While the connection between Vcc and an input (or output) is the preferred method for using a pull-up resistor, the question arises as how do we calculate the value of the resistance require to ensure the correct operation of the input.

Calculating Pull-up Resistor Value

All digital logic gates, circuits and micro-controllers are limited not only by their operating voltage, but in the current sinking and sourcing ability of each input pin. Digital logic circuits operate using two binary states which are normally represented by two distinct voltages: a high voltage V_H for logic "1" and low voltage V_L for logic "0". But within each of these two voltage states, there is a range of voltages which define the upper and lower voltages of these two binary states.

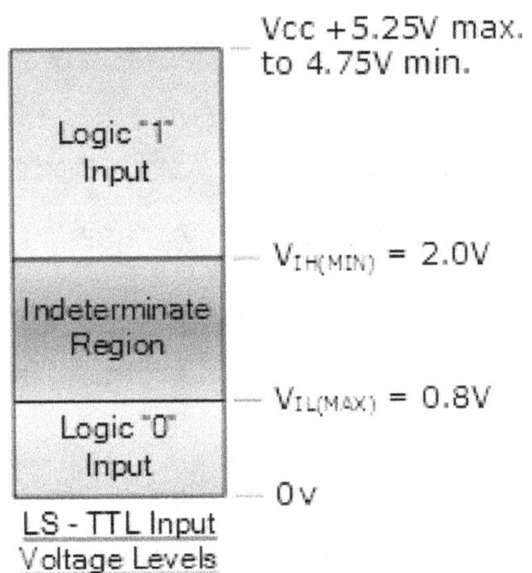

Therefore, for example, for the TTL 74LSxxx series of digital logic gates, the voltage ranges representing a logic level "1" and a logic level "0" are shown.

Where: $V_{IH(min)}$ = 2.0V is the minimum input voltage guaranteed to be recognized as a logic "1" (high) input and $V_{IL(max)}$ = 0.8V is the maximum input voltage guaranteed to be recognized as a logic "0" (low) input.

In other words, TTL 74LSxxx input signals between 0 and 0.8V are considered "LOW", and input signals between 2.0 and 5.0V are considered "HIGH". Any voltage in between 0.8 and 2.0 volts is not recognized as a logic "1" or logic "0".

When logic gates are connected together, the current flows between the output of one logic gate and the input of another. The amount of current required by a basic TTL logic gate input depends on whether the input is a logic "0" (LOW) or a logic "1" (HIGH) as this creates a current-Sourcing action for a logic "0" and a current-sinking action for a logic "1".

When the input of the logic gate is HIGH, a current flows into the TTL input as the input acts basically as a path connected directly to ground. This input current, $I_{IH(max)}$ is positive in value as it flows "into" the gate and for most TTL 74LSxxx inputs have a value of 20µA.

Likewise, when the input of the logic gate is LOW, the current flows out of the TTL input as the input acts basically as a path connected directly to Vcc. This input current, $I_{IL(max)}$ is negative in value as it flows "out-of" the gate and for most TTL 74LSxxx inputs, has a value of -400µA, (-0.4mA).

Note that the values of HIGH and LOW voltages and currents differ between TTL logic families and is Also much, much lower for CMOS logic families. Also, the input voltage and current requirements for micro-controllers, PIC, Arduino, Raspberry Pie, etc. will Also be different Therefore please consult their data sheets first.

By knowing the information above, we can calculate the maximum pull-up resistor value required for a single TTL 74LS series logic gate as:

Single Gate Pull-up Resistor Value

$$R_{MAX} = \frac{V_{CC} - V_{IH(MIN)}}{I_{IH}} = \frac{5 - 2}{20 \times 10^{-6}} = 150K\Omega$$

Then using Ohms Law, the maximum pull-up resistance required to drop 3 volts for a single TTL 74LS series logic gate would be 150kΩ. While this calculated value would work, it leaves no room for error as the voltage drop across the resistor is at its maximum while the input current is at its minimum.

Ideally, we would want a logic "1" to be as close to Vcc as possible to guarantee 100% that the gate sees a HIGH (logic-1) input through the pull-up resistor. Reducing the resistive value of this pull-up resistor would give us a greater error margin should the tolerance of the resistor or the supply voltage not be as calculated. However, we do not want the resistor value to be too low as this would increases current flow into the gate increasing power dissipation.

Therefore, if we assume a voltage drop of only one volt, (1.0V) across the resistor giving double the input voltage at 4 volts, a quick calculation would give us a single pull-up resistor value of 50kΩ. Reducing the resistive value further, will produce a smaller voltage drop but increase the current.

It has been observed that while there may be a maximum allowable resistive value, the resistance value for pull-up resistors is not usually that critical with resistance values ranging from between 10k to 100k ohms acceptable.

This simple example above provides us the maximum value of the pull-up resistor required to bias a single TTL gate. But we can Also use the same resistor to bias multiple inputs to a logic "1" value.

For example, let's assume we have constructed a digital circuit and that there are ten unused logic gate inputs. As a single standard TTL 74LS gate, has an input current, $I_{IH(max)}$ of 20µA (Also called a fan-in of 1), then ten TTL logic gates would require a total current of: 10 x 20µA = 200µA representing a fan-in of 10. Therefore, the maximum resistive value of the pull-up resistor required to supply ten unused inputs would be calculated as follows:

Multiple Gate Pull-up Resistor Value

$$R_{MAX} = \frac{V_{CC} - V_{IH(MIN)}}{10 \times I_{IH}} = \frac{5 - 2}{10 \times 20 \times 10^{-6}} = 15K\Omega$$

Here the fan-in is given as 10, but if "n" TTL inputs are connected together then the current through the resistance would be "n" times $I_{IH(max)}$. Again, as before, this 15kΩ resistance may be the exact calculated value, but leaves no room for error Therefore reducing the voltage drop to one volt (or any value you want) gives a resistive value of only 5kΩ.

Pull-up Resistor Example No1

Two TTL 74LS00 NAND Gates along with a single-pole double-throw switch are to be used to make a simple Set-Rest bistable flip-flop. Calculate: 1). The maximum pull-up resistors values if the voltage representing a logic HIGH input is to be held at 4.5 volts when the switch is open, and 2). The current flowing through the resistor when the switch is closed (assume zero contact resistance). Also draw the circuit.

Data given: Vcc = 5V, V_{IH} = 4.5V, and $I_{IH(max)}$ = 20µA

1). Pull-up Resistors value, R_{MAX}

$$R_{MAX} = \frac{V_{CC} - V_{IH}}{I_{IH}} = \frac{5 - 4.5}{20 \times 10^{-6}} = 25K\Omega$$

2). Resistor Current, I_R

$$I_R = \frac{V_{CC}}{R} = \frac{5V}{25k\Omega} = 200\mu A \text{ or } 0.2mA$$

Set-Reset Bistable Circuit

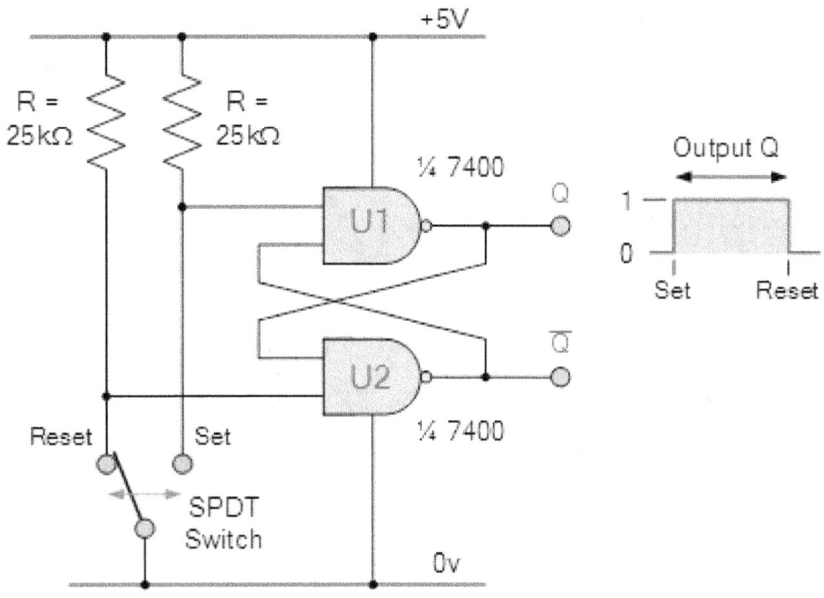

Pull-down Resistors

A *Pull-down resistor* works in the same way as the previous pull-up resistor, except this time the logic gates input is tied to ground, logic level "0" (LOW) or it may go HIGH by the operation of a mechanical switch.

This pull-down resistor configuration is particularly useful for digital circuits like latches, counters and flip-flops that require a positive one-shot trigger when a switch is momentarily closed to cause a state change.

While they may seem to operate in the same way as the pull-up resistor, the resistive value of a passive pull-down resistor is more critical with TTL logic gates than with similar CMOS gates. This is As a TTL input Sources much more current out of its input in its LOW state.

From above we saw that the maximum voltage level that represents a logic "0" (low) for a TTL 74LSxxx series logic gate is between 0 and 0.8 volts, ($V_{IL(MAX)}$ = 0.8V). Also, when LOW, the gate Sources current to the value of 400µA, (I_{IL} = 400µA). The maximum pull-down resistor value for a single TTL logic gate is therefore calculated as:

Single Gate Pull-down Resistor Value

$$R_{MAX} = \frac{V_{IL(MAX)} - 0}{I_{IL}} = \frac{0.8 - 0}{400 \times 10^{-6}} = 2K\Omega$$

Then the maximum pull-down resistor value is calculated as 2kΩ. Again, as with the pull-up resistor calculations, this 2kΩ resistor value leaves no room for error as the voltage drop is at maximum.

Therefore, if the resistance is too large, the voltage drop across the pull-down resistor may result in a gate input voltage beyond the normal LOW voltage range, therefore to ensure correct switching it is better to have an input voltage of 0.5 volts or less.

Pull-down Resistor Application

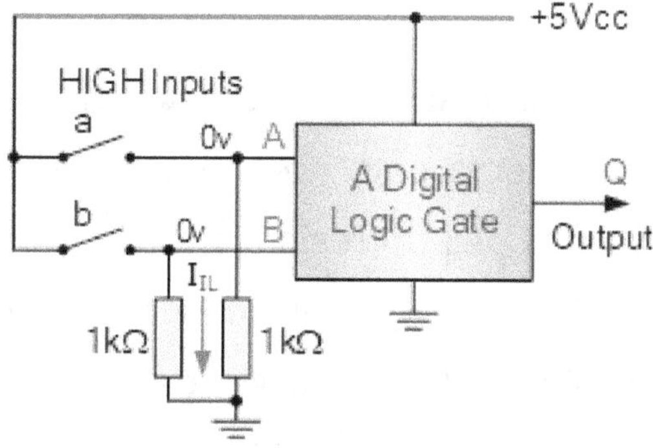

Therefore, if we assume a voltage drop of only 0.4 volts across the resistor, a quick calculation would give us a single pull-down resistor value of 1kΩ.

Reducing the resistive value further, will produce a smaller voltage drop tying the input further to ground (low). This datasheet value of 400µA or 0.4mA (I_{IL}) is the minimum LOW current value but it may be higher.

Also, connecting inputs together will result in a larger current through the resistor. For example, a fan-in of 10 will result in $10 \times 400\mu A = 4.0mA$ requiring a pull-down resistance of 100Ω.

But you might be thinking, why use a pull-down resistor at all when a direct connection to ground (0V) would produce the required LOW? A direct connection to ground without the pull-down resistor would certainly work in most cases, but as the gates input is permanently tied to ground, the use of a resistor limits the current flowing out of the input thereby reducing power loss while still maintaining a logic "0" condition.

Open-collector Outputs

Thus far we have seen that we can use either a pull-up resistor or a pull-down resistor to control the voltage level of a logic gate. But we can Also use pull-up resistors on the output of a gate to allow different gate technologies to be connected, for example TTL to CMOS or for transmission line driving applications that require higher currents and voltages.

In order to overcome this some logic gates are manufactured with the collector of the gates internal output circuitry left open meaning that the logic gate does not actually drive the output HIGH, only LOW as its the job of external pull-up resistors to do this. One example of this is the TTL 74LS01, Quad 2-input NAND gate which has open collector outputs, as opposed the standard TTL 74LS00, Quad 2-input NAND gate.

Open-collector, (OC) or open-drain for CMOS, outputs are Normally used in buffer/inverter/driver ICs (TTL 74LS06, 74LS07) allowing for a greater output current and/or voltage capability than you would get with ordinary logic gates.

For example, to drive a large load such as an LED indicator, a small relay or dc motor. Either way, the principle and use of the pull-up resistor is pretty much the same as for the input.

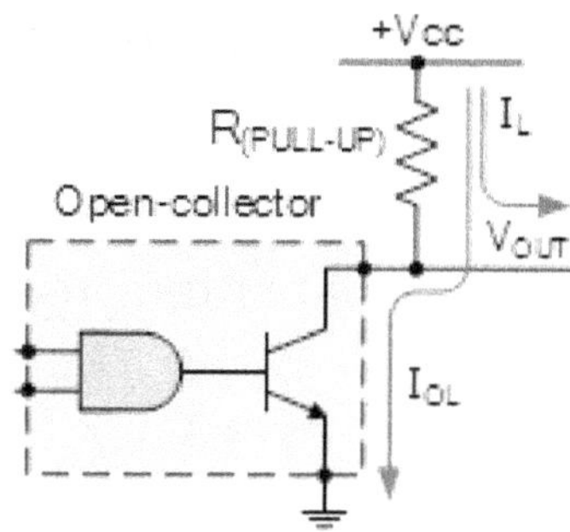

Logic gates, micro-controllers and other such digital circuits that have open-collector outputs, are incapable of pulling their outputs HIGH as there is no internal path to the supply voltage, (Vcc). This condition means that their output is either grounded when LOW, or floating when HIGH, therefore an external pull-up resistor, (Rp) needs to be connected from the open-collector terminal of the pull-down transistor to the Vcc supply.

With pull-up resistors connected, the output still works in the same way as a normal logic gate in that when the output transistor is OFF (open), the output is HIGH, and when the transistor is ON (closed), the output is LOW. Thus, the transistor turns ON to pull the output to a LOW level.

The size of the pull-up resistor depends on the connected load and the voltage drop across the resistor when the transistor is OFF. When the output is LOW, the transistor must be able to sink the load current through the pull-up resistor. Likewise, when the output is HIGH, the current through the pull-up resistor must be high enough for whatever is connected to it.

As we saw before with the input, the output of a digital logic gate operates using two binary states which are represented by two distinct voltages: a high voltage V_H for logic "1" and low voltage V_L for logic "0". Within each of these two voltage states, there is a range of voltages which define their upper and lower voltages.

$V_{OH(min)}$ is the minimum output voltage guaranteed to be recognized as a logic "1" (HIGH) output and for TTL this is given at 2.7 volts. $V_{OL(max)}$ is the maximum output voltage guaranteed to be recognized as a logic "0" (LOW) output and for TTL this is given as 0.5 volts. In other words, TTL 74LSxxx output voltages between 0 and 0.5V are considered "LOW", and output voltages between 2.7 and 5.0V are considered "HIGH".

Therefore, when using open-collector logic gates, the pull-up resistors value required can be determined from the following equation:

Open-collector Pull-up Resistors Value

$$R_{MIN} = \frac{V_{CC} - V_{OL(MAX)}}{I_{OL}} = \frac{5 - 0.5}{8 \times 10^{-3}} = 562\Omega$$

Where the values for a 7401 open-collector NAND are given as: Vcc = 5V, V_{OL} = 0.5V, and $I_{OL(max)}$ = 8mA. Note that it is important to calculate a suitable pull-up resistor Rp as the current through the resistor must not exceed $I_{OL(max)}$.

We said earlier that open-collector logic gates are ideal for driving loads that require higher voltage and current levels, such as an LED indicator. The TTL 74LS06 Hex Inverter Buffer/Driver has an $I_{OL(max)}$ rating of 40 mA (instead of 8mA for the 74LS01) and a $V_{OH(max)}$ rating of 30 volts instead of the usual 5 volts (but the IC itself MUST use a 5V supply). Then the 74LS06 will allow us to drive a load up to 40mA of current.

Pull-up Resistors Example No2

A 74LS06 Hex Inverter Driver is required to control a single red LED indicator from a 12 volt supply. If the LED requires 15mA at 1.7V voltage drop and the V_{OL} of the HEX Inverter when fully ON is 0.1 volts, calculate the value of the current limiting resistor required to drive the LED.

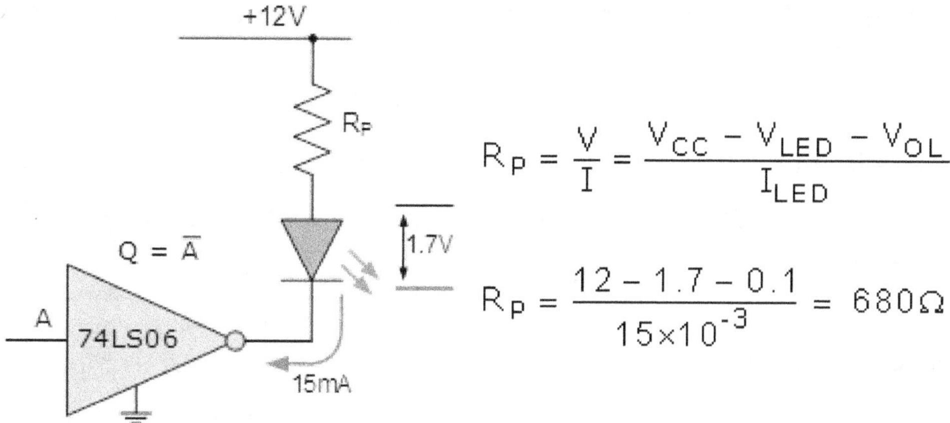

$$R_P = \frac{V}{I} = \frac{V_{CC} - V_{LED} - V_{OL}}{I_{LED}}$$

$$R_P = \frac{12 - 1.7 - 0.1}{15 \times 10^{-3}} = 680\,\Omega$$

We can use open-collector drivers in a similar way to drive small electromechanical relays, lamps or dc motors as these devices typically require 5V or 12V or more, at a current of about 10 to 20 mA's to operate correctly.

Two or more open-collector outputs of TTL gates can be directly connected together and tied through a single external pull-up resistor. The result is that the outputs are effectively AND 'ed together as the combination behaves as if the gates were connected to an AND gate. This type of configuration is called wired AND logic.

Pull-up Resistors in A Brief

We have seen here in this chapter about passive pull-up and pull-down resistors that when left open–circuited, the inputs of digital logic gates may self–bias or float about to whatever logic level they choose and many switching errors can be traced back to unconnected and floating input pins.

Pull-up resistors connect unused input pins (AND and NAND gates) to the dc supply voltage, (Vcc) to keep the given input HIGH. A pull-down resistor connects unused input pins (OR and NOR gates) to ground, (0V) to keep the given input LOW. The resistance value for a pull-up resistor is not usually that critical but must maintain the input pin voltage above V_{IH}. The use of 10kΩ pull-up resistor is common but values can range from 1k to 100k ohms.

Pull-down resistors are a little more critical As of the low input voltage level, $V_{IL(max)}$ and the higher I_{IL} current. The use of 100Ω pull-down resistor is the most common but they can range in resistive value from 50 up to 1k ohms.

Digital logic gates with open-collector (in the case of the TTL logic) outputs or open-drain (in the case of the CMOS logic) outputs need to connect to an external pull-up resistor between their output pin and the dc power supply to make the logic gate perform the intended logic function.

The advantage of using open collector/open drain gates is in their capability to switch higher voltages and currents or their ability of provide wired ANDing operation. Therefore, open-collector gates like the 74LS06 are capable of driving larger loads since their outputs can be connected to supplies of up to 30 volts via an external pull-up resistor.

Universal Logic gates can be used to produce any other logic or Boolean function with the NAND and NOR gates being minimal

Individual logic gates can be connected together to form a variety of different switching functions and combinational logic circuits. Since, three most basic logic gates are the: AND, OR and NOT gates, and given this set of logic gates it is possible to implement all of the possible Boolean switching functions, thus making them a "full set" of **Universal Logic Gates**.

By using logical sets in this way, the various laws and theorems of Boolean Algebra can be implemented with a complete set of logic gates. In fact, it is possible to produce every other Boolean function using just the set of AND and NOT gates since the OR function can be created using just these two gates. Likewise, the set of OR and NOT can be used to create the AND function.

Any logic gate which can be combined into a set to understand all other logical functions is said to be a universal gate with a complete logic set being a group of gates that can be used to form any other logic function.

For example, AND and NOT constitute a complete set of logic, as does OR and NOT as cascading together an AND with a NOT gate would give us a NAND gate. Similarly cascading an OR and NOT gate together will produce a NOR gate, and so on. However, the two functions of AND and OR on their own do not form a complete logic set.

Therefore, by using these three *Universal Logic Gates* we can create a range of other Boolean functions and gates. However, the NAND and NOR gates are classed as minimal sets as they have the property of being a complete set in themselves since they can be used individually or together to construct many other logic circuits. Therefore, we can define the complete sets of operations of the main logic gates as follows:

- AND, OR and NOT (a Full Set)
- AND and NOT (a Complete Set)
- OR and NOT (a Complete Set)
- NAND (a Minimal Set)
- NOR (a Minimal Set)

Thus, we can use these five sets of gates, together or individually as the building blocks to produce more complex logic circuits called *combinational logic circuits*. But first let us remind ourselves of the switching characteristics of the three basic logic gates, AND, OR and NOT.

The AND Function

In mathematics, the number or quantity obtained by multiplying two (or more) numbers together is called the *product*. In Boolean Algebra the AND function is the equivalent of multiplication. As a result, its output state represents the product of its inputs. The AND function is represented in Boolean Algebra by a single "dot" (.) Therefore for a two input AND gate the Boolean equation is given as: Q = A.B, that is Q equals both A AND B.

The 2-input Logic AND Gate

Symbol

Truth Table

B	A	Q
0	0	0
0	1	0
1	0	0
1	1	1

The OR Function

In mathematics, the number or quantity obtained by adding two (or more) numbers together is called the *sum*. In Boolean Algebra the OR function is the equivalent of addition Therefore its output state represents the addition of its inputs. In Boolean Algebra the OR function is represented by a "plus" sign (+) Therefore for a two input OR gate the Boolean equation is given as: Q = A+B, that is Q equals either A OR B.

The 2-input Logic OR Gate

Symbol

Truth Table

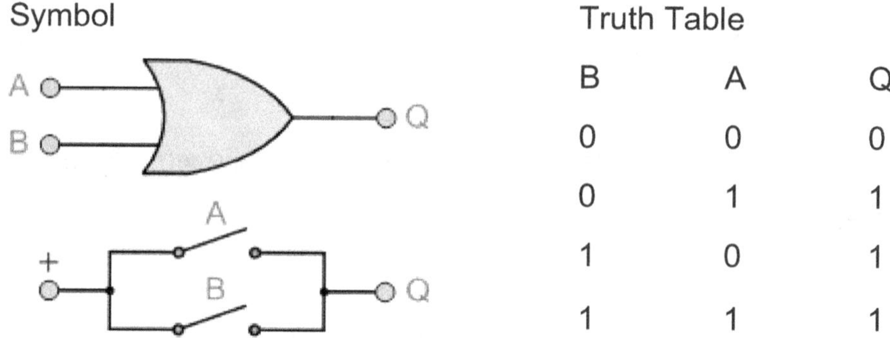

B	A	Q
0	0	0
0	1	1
1	0	1
1	1	1

The NOT Function

The NOT gate, which is known as an "inverter" is given a symbol whose shape is that of a triangle pointing to the right with a circle at its end. This circle is known as an "inversion bubble".

The NOT function is not a decision-making logic gate like the AND, or OR gates, but instead is used to invert or complement a digital signal. In other words, its output state will always be the opposite of its input state.

The NOT gate symbol has a single input and a single output as shown.

The Logic NOT Gate

Symbol

Truth Table

A	Q
0	1
1	0

The single input NOT gate or invert function can be cascaded with itself to produce what is called a digital buffer. The first NOT gate will invert the input and the second will re-invert it back to its original level performing a double inversion of the single input. Non-inverting Digital Buffers have many uses in digital electronics as this double inversion of the input can be used to provide digital amplification and isolation of the circuit.

Using the AND and NOT Set

Using just the AND and NOT set of logic gates we can create the following Boolean functions and equivalent gates.

Universal Logic Gates AND/NOT Set Equivalents

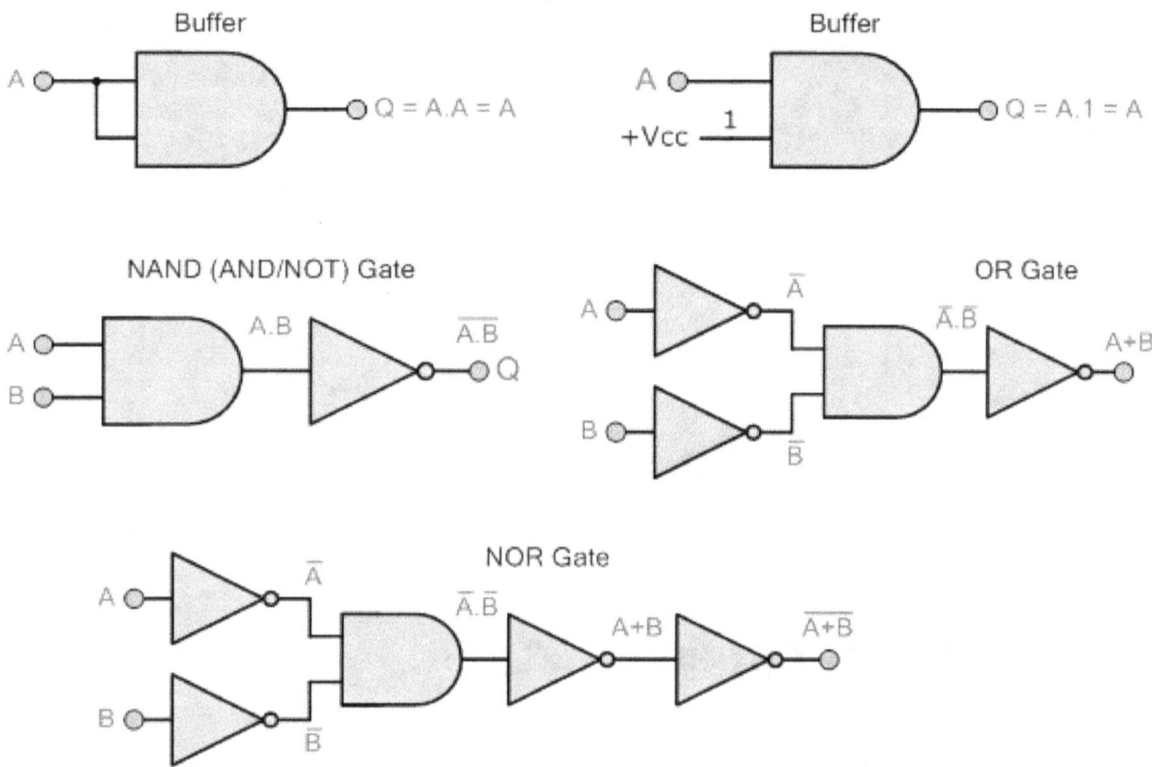

Using the OR and NOT Set

Using the OR and NOT set of logic gates we can create the following Boolean functions and equivalent gates.

OR/NOT Set Equivalents

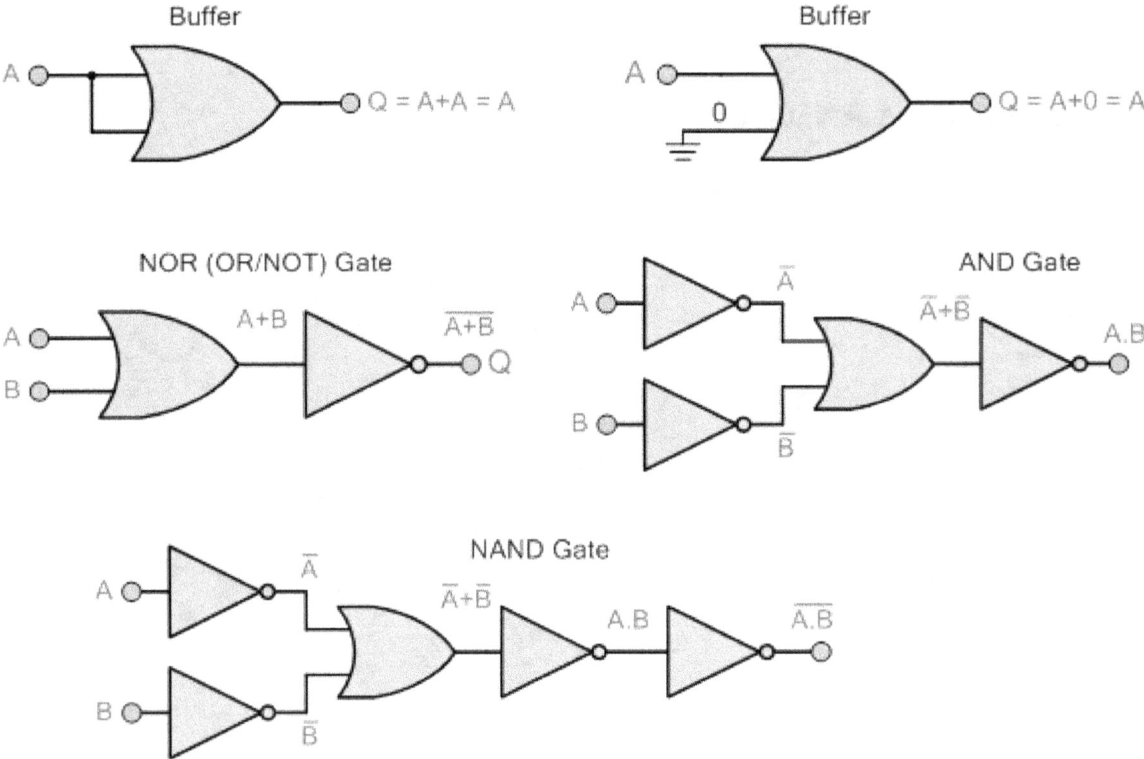

Using the Full AND, OR and NOT Set

Using the full AND, OR and NOT set of logic gates we can create the Boolean expressions for the Exclusive-OR (Ex-OR) and the NOT Exclusive-OR (Ex-NOR) gates as shown.

Full AND/OR/NOT Set to Implement Ex-OR

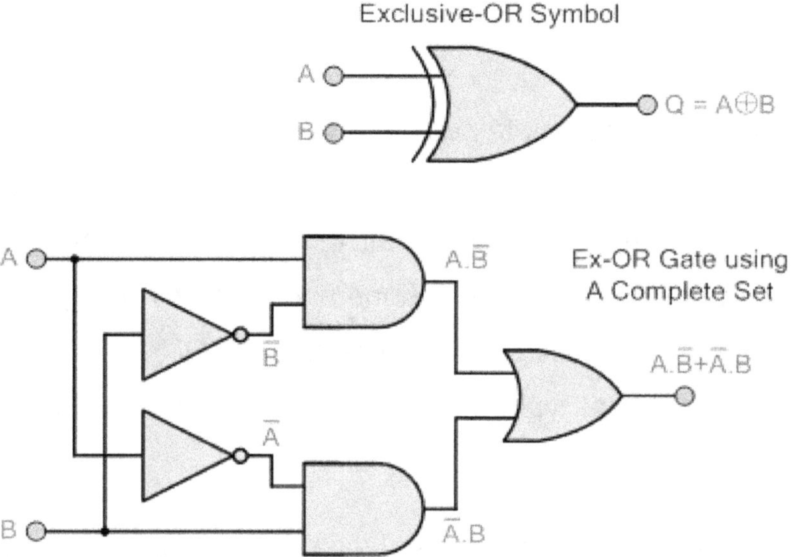

Full AND/OR/NOT Set to Implement Ex-NOR

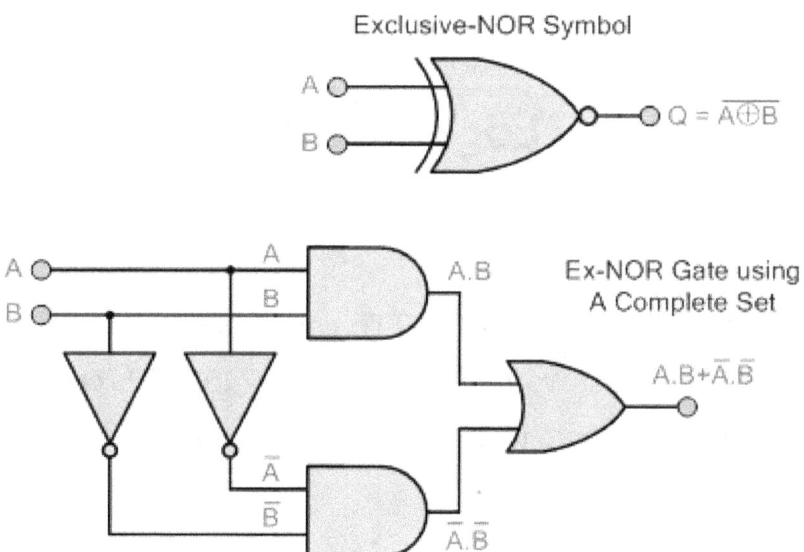

Note that neither the Exclusive-OR gate or the Exclusive-NOR gate can be classed as a universal logic gate as they cannot be used on their own or together to produce any other Boolean function.

Universal Logic Gates

One of the main disadvantages of using the complete sets of AND, OR and NOT gates is that to produce any equivalent logic gate or function we require two (or more) different types of logic gate, AND and NOT, or OR and NOT, or all three as shown above. However, we can understand all of the other Boolean functions and gates by using just one single type of universal logic gate, the NAND (NOT AND) or the NOR (NOT OR) gate, thereby reducing the number of different types of logic gates required, and also the cost.

The NAND and NOR gates are the complements of the previous AND and OR functions respectively and are individually a complete set of logic as they can be used to implement any other Boolean function or gate. But as we can construct other logic switching functions using just these gates on their own, they are both called a minimal set of gates. Thus, the NAND and the NOR gates are Normally referred to as **Universal Logic Gates**.

Implementation of Universal Logic Gates Function Using Only NAND

The 7400 (or the 74LS00 or 74HC00) quad 2-input NAND TTL chip has four individual NAND gates within a single IC package. Thus, we can use a single 7400 TTL chip to produce all the Boolean functions from a NOT gate to a NOR gate as shown.

Universal Logic Gates using only NAND Gates

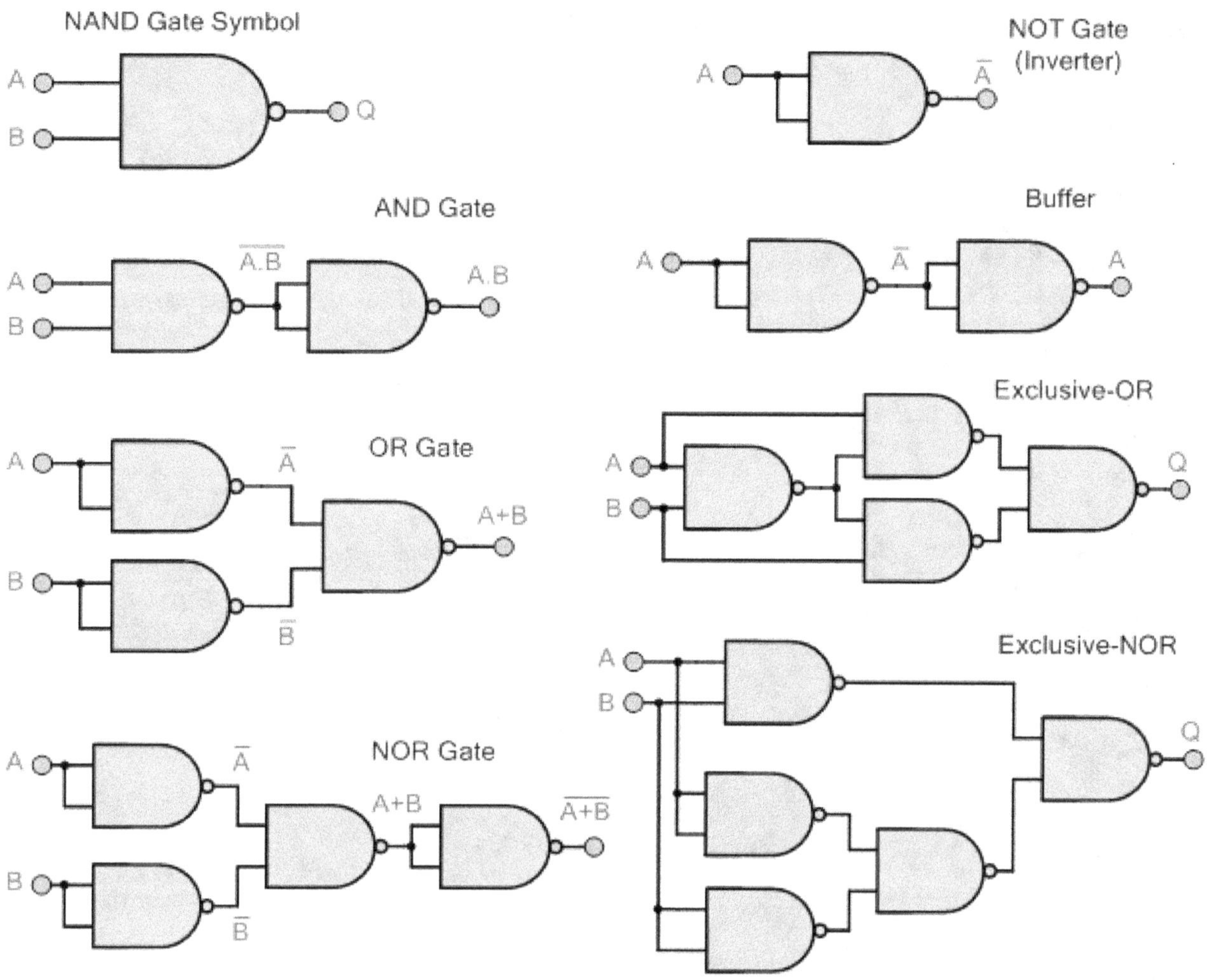

In this way, ALL other logic gate functions can be created using only NAND gates making it a universal logic gate.

Implementation of Universal Logic Gates Function Using Only NOR

The 7402 (or the 74LS02 or 74HC02) quad 2-input NOR TTL chip has four individual NOR gates within a single IC package. Thus like the previous 7400 NAND IC we can use a single 7402 TTL chip to produce all the Boolean functions from a single NOT gate to a NAND gate as shown.

Universal Logic Gates using only NOR Gates

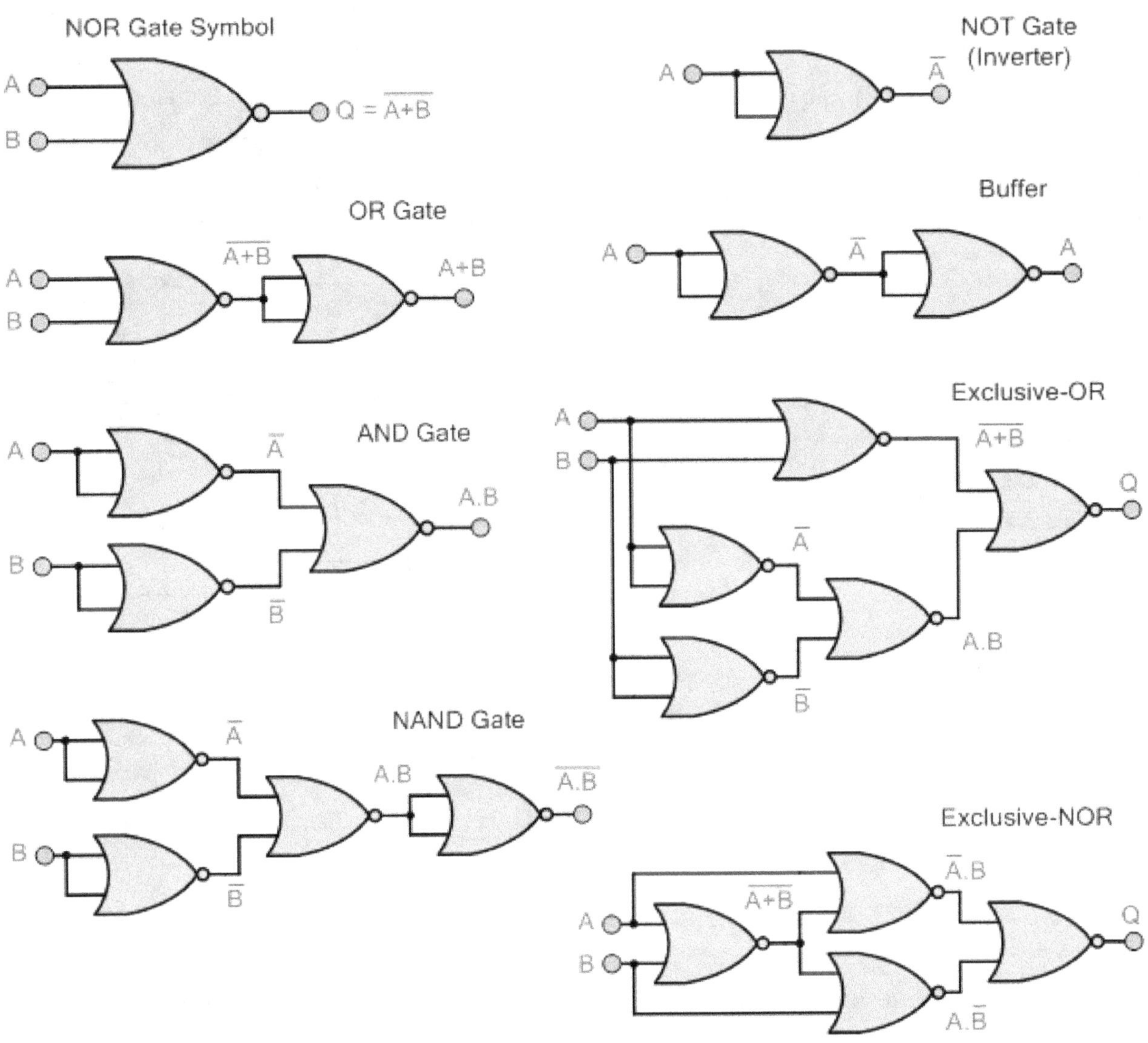

In this way, all other logic gate functions can be created using only NOR gates making it a universal logic gate.

It should be noted here that the implementation of the Exclusive-OR gate is more efficient using NAND gates compared to using NOR gates, while the implementation of the Exclusive-NOR gate is more efficient with NOR gates compared to using NAND gates as in each case only four individual logic gates are required. In other words, we can create all the Boolean functions using just one 7400 NAND or one 7402 NOR chip including its various sub-families.